PECULIAR DEATHS OF FAMOUS MATHEMATICIANS

PECULIAR DEATHS OF FAMOUS MATHEMATICIANS

IOANNA GEORGIOU

Illustrations by Asuka Young

Publisher's Note

I should make a confession: I am a historian by training. No apologies are necessary, it's just that people expect the Publisher at Tarquin to be a mathematician. History, though, is the story of everything past – and that includes the history of mathematics. It's a fascinating story too – uniquely a story of progress. One truth developed from the truths discovered or found by previous generations. And because of the unique nature of proof, less subject to side tracking by belief systems than most disciplines including the other sciences.

So the history of mathematics is one of development – as Newton put it, of individuals "standing on the shoulders of giants". You, dear reader, are about to read about some of the stories of those giants – maybe they will help you on your journey to stand on the shoulders of giants.

But, whatever you get from it, have some fun working out which death is NOT true. And see if you can find 72 anachronisms in the pictures as you go through!

ISBN (book): 9781913565701
ISBN (ebook): 9781913565718

Design by Karl Hunt

Printed in the UK

Published by Tarquin
Suite 74, 17 Holywell Hill
St Albans AL1 1DT
United Kingdom
info@tarquingroup.com
www.tarquingroup.com

CONTENTS

INTRODUCTION

Mathematicians may have to live a bit more in their heads than most other professionals, save perhaps authors and poets, as that is where most of the magic happens.

Mathematics created or invented (whichever philosophical inclination you follow) often precedes any possible practical applications. Perhaps because of the lack of immediate apparent application of their results, mathematicians are often misunderstood as solitary, unemotional, or just bizarre people. But the mathematical "magic" does not happen overnight; it is usually work in progress like many other professional occupations. In the meantime, mathematicians get on with living their lives. Like anyone else they maintain a circle of friends, raise families, grow their professional networks and so on. If they are lucky enough to stumble upon something great, they may get a theorem or two named after them.

When the time comes for them to depart, some do so more impressively than others.

In this book we will be telling the life stories of ten mathematicians and take a look at their mathematical worlds, and their findings. Please do bear in mind that the older the story the harder it is to confirm certain details. Here, we go by what seems to have survived throughout the centuries – this is what we will take as "true". The glue that holds these stories together – apart from mathematics – is how they have died in the most peculiar ways.

But there's a catch!

Only nine out of the ten mathematicians we talk about in this book have died a peculiar death. One of those ten is a figment of my imagination. Your challenge is to work out which one!

At the end of each chapter, I present you with statistical information about the likelihood of each of those circumstances and deaths. Note that "negligible" would mean there was substantial difficulty in identifying more than one such case. It is then up to you to make an informed guess. The truth is revealed in the final chapter, where any injustices committed will be acknowledged.

As an additional intellectual snack, most of the illustrations contain an anachronism; an item or symbol misplaced in time. Can you spot the items in the pictures that are ahead of their time? See the notes in the end of the book for answers – see if you can find them all!

Special thanks to the colleagues who chatted with me about this book, shared their ideas, stories and insights.

CAN BEANS KILL YOU?

PYTHAGORAS

(C.570-495 BC)

PYTHAGORAS'S LIFE

Pythagoras lived in the 6th century BC. He was born in Samos, a Greek island in the Aegean Sea. Samos was home to other scientists, engineers and philosophers of the time. Pythagoras is by far the best known, however, and in 1955 one of the towns on the island was renamed Pythagoreio. Before then it was called Tigani, which is the Greek word for "pan", possibly due to the shape of the port. I am sure we can agree the more recent option is far better. At that same port there is a special statue to honour Pythagoras, and his namesake theorem, as the formation includes a right-angle triangle.

For the Greeks, the known world was anything surrounding the Mediterranean, the literal meaning of the word being *middle of the earth*. The sea *was* easily accessible by ship, scholars, merchants, scientists and many others would travel around, learning from one another, trading, and discovering new habits, practices and materials. Pythagoras was a keen traveller, having visited Egypt, Persia and eventually settling in Italy. He pursued a

variety of interests, ranging widely from mathematics to astronomy, music and pottery.

One of his very impressive inventions, albeit perhaps not the most famous one, was the Pythagorean cup (or fair cup). This pottery invention assured that guests around a table were served wine fairly, with no one getting bigger portions. This unique invention, still reproduced today in small local potteries, functions on a principle similar to the straw (plastic or paper one but before it melts in your drink).

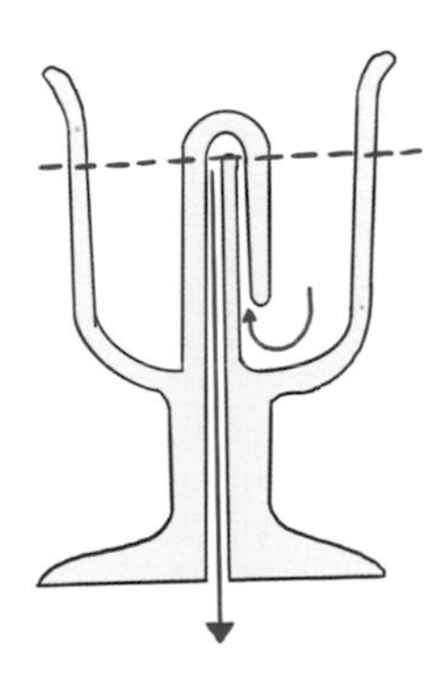

When we drink through a straw, essentially, we remove the air from the straw tube, and the atmospheric pressure acting on the surface of the drink pushes it upwards through the straw. But we need to continue to withdraw the liquid or the air, otherwise the flow of the drink through the straw stops, as it is filled with air again and there is a balance of pressure.

But what would happen if the end of the bent straw was lower than the level of the drink? In that case, we would not be able to stop the flow of the drink through the straw and we would lose all of the liquid. Try it at home with water; it is not dangerous unless you spill it on the floor and you might slip.

The Pythagorean cup includes an inner tube, hidden with a weird protrusion. The hollow tube continues to the bottom of the cup. By pouring liquid that exceeds the line marked, the top of the tube is reached, and the liquid pours out through the hole as the air from the inside tube is removed. The end part is obviously below the level of the drink, and the air cannot be restored in that tube. Hence the atmospheric pressure continues to do its magic, eventually leaving the cup empty for the greedy ones.

You might wonder though, why was Pythagoras so concerned with sharing wine fairly? This ideal was part of his teaching within the school (or brotherhood) he founded, called the Pythagoreans. Pythagoras's followers, a bit like disciples, adhered to specific rules and followed a special lifestyle. They were studying daily, observing silence – up to and including being sworn to secrecy – and were rather unwilling to share their findings with the general public. They made progress in areas of mathematics, sciences, music, and more. They also believed that "all is number"; that everything in the world can be described as whole numbers or ratios between two numbers.

PYTHAGORAS'S MATHEMATICS

The thing Pythagoras is most famous for is the theorem that bears his name. The square of the hypotenuse is equal to the sum of the squares of the two perpendicular sides.

Did Pythagoras discover this right-angle triangle theorem? Is this why it bears his name? It seems that this theorem was named after Pythagoras by Euclid[1], in his mathematical treatise "Elements", consisting of 13 books, approximately 200 years *after* Pythagoras' death. No one else seems to have called it that, not even the Pythagoreans, until Euclid did in Book I proposition number 47.

Many Greeks of his era were sceptical about the relatively new technology of writing, which was regarded dangerous and novel even by Socrates.[2] Pythagoras like many Greeks preferred oral or non-written transition of knowledge. In terms of attribution of his work, this creates a problem, since there is not much information about him written during his lifetime. Additionally, the practicalities of this theorem were in use for up to 1000 years before Pythagoras was even born! The evidence is carved on some Babylonian clay tablets. This includes diagrams and the number combinations (nowadays called Pythagorean triples). And Egyptians were using knotted ropes in the ratio 3:4:5 as a right-angle checking device, possibly useful for constructions.

1 Greek mathematician mid-4th century BC
2 Greek philosopher 4th century BC

However, proof was of the utmost importance in Ancient Greece, where mathematics progressed beyond its practical uses that earlier civilisations were focused upon. Indeed, as we have said, for Pythagoras and his followers, numbers underlie everything – mathematics is an essential part of their quasi-religious world view.

So, maybe, his major contribution was to have proved the theorem and hence it was named after him? We can only speculate. However, there are surviving versions of a legend that upon proving the theorem, Pythagoras sacrificed (up to) 100 oxen to honour the gods. A fact which is inconsistent with him and the Pythagoreans being strict vegetarians, and believing in reincarnation . . . so scepticism is probably justified.

But let's have a look at one of the many (over 300) proofs of the theorem.

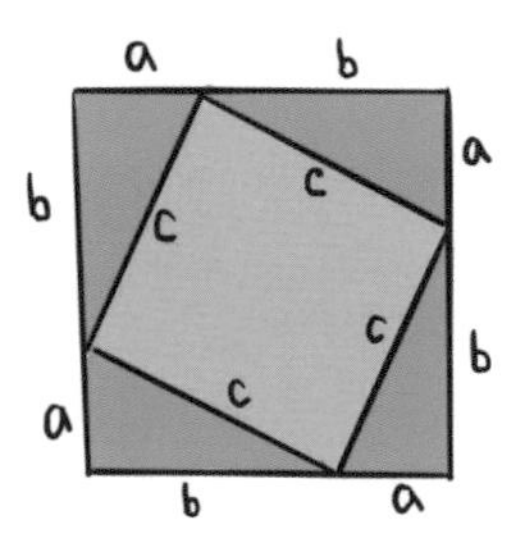

Start with a square of side $a + b$. It can be seen as a large square, or as a smaller square surrounded by four smaller right-angle triangles. But of course, however it is seen, it will have a constant area.

So, large square area:

$$(a + b)^2 = a^2 + 2ab + b^2.$$

Smaller square plus triangles area:

$$c^2 + 4 \times (\tfrac{ab}{2}) = c^2 + 2ab.$$

Since the two expressions describe the same thing hence equal to one another:

$a^2 + 2ab + b^2 = c^2 + 2ab$ remove $2ab$ from both sides and we get:
$a^2 + b^2 = c^2$, Pythagoras' theorem!

THE DEATH OF PYTHAGORAS

Pythagoras ended up in Croton in modern day Italy after much travelling and many adventures. There, the Pythagoreans became a powerful cult and extremely selective of their membership. The son of a nobleman, used to special treatment, was refused entry to the brotherhood and was greatly upset by what he perceived as a slight. He used his wealth and influence to turn the Crotonates against the Pythagoreans. So much so that they set the Pythagoreans' homes on fire and chased them with knifes. Pythagoras, with the help of his faithful followers, got a good head start and seemed to have a high chance of surviving the attack by the angry mob. In his attempt to escape his pursuers, he continued to run, until he reached the edge to a huge broad bean field. This would seem like good luck for most people. With their long stems, the broad bean plants could have offered Pythagoras the perfect place to hide until his pursuers would have given up!

But it was not quite as simple for Pythagoras, who avoided broad beans at all costs (literally). His sect thought that a curse would be inflicted on them by broad beans. We can only speculate as to the reasons why they developed this aversion. Perhaps it could be that the peculiar shape of a broad bean resembles a human foetus, and that the flesh of the bean to be too meaty for a vegetarian cult? As we know today, if consumed raw, broad beans can cause a haemolytic response which could be a more reasonable explanation for avoiding eating them. But how about avoiding going near them?

Legend has it Pythagoras refused to enter the field that could prove his salvation. His superstition against the beans was too strong to overcome the threat of death and he was stubbed to death just at the edge of the field. One cannot help but wonder, how the curse could have been worse than his fate?

AND NOW SOME STATISTICS

Percentage of deaths by stabbing: 2% (most stabbings are for robberies and other disputes)

Percentage of deaths by broad bean consumption: 8% of those with favism (genetic disorder)

Percentage of deaths by approaching a broad bean field: negligible

---

DEATH BY SQUARE ROOT

HIPPASUS

(C.530-450 BC)

HIPPASUS'S LIFE

Hippasus was born around 530BC in Metapontum, a city in Magna Graecia, the area in Southern Italy where Greek was the predominant language at the time. There is little known about his life before he became a member of the Pythagoreans. Much of what was written about him came long after his death. Being a Pythagorean meant he was devoted to studying and adhered to the principles of life within the brotherhood. His studies included philosophy, music and of course, mathematics. The Pythagoreans valued knowledge above most other things, and only spoke if they thought what they had to say would improve on silence. Perhaps that would be a good rule for most people.

As a philosopher he spent much of his time contemplating cosmogonic issues – the origins of the universe. He considered fire to be the building element of the universe. That was different than what other Pythagoreans contended, in that fire is an element (or state) found in nature, and that the building blocks were numbers. One could say he was more pragmatic than his colleagues, who found metaphysical entities more plausible starts to the universe.

Another area Hippasus must have worked extensively on was the theory of sound. Music was amongst the Pythagorean interests. Acoustics relied extensively on physical experiments. Hippasus created four metal discs of the same diameter but of increasing thickness. The second was 1⅓ as thick as the first, the third was 1½ as the first and the fourth to be double the thickness of the first. By striking the discs in any combinations of two, he demonstrated that the same harmonic sounds were produced as if strings were used. The feature that had to be kept constant was the ratio between the thickness of the discs or the length of the strings. Those ratios would result in harmonic intervals, a term from music theory, indicating pleasing combinations of notes. For example, comparing the disc that was 1.5 times thicker than the first would create a ratio of 3:2. That would result in a "just fifth" sound in music theory terms. In other words, by raising the frequency produced by a factor of 1.5, one can create the pleasing sound of a "just fifth".

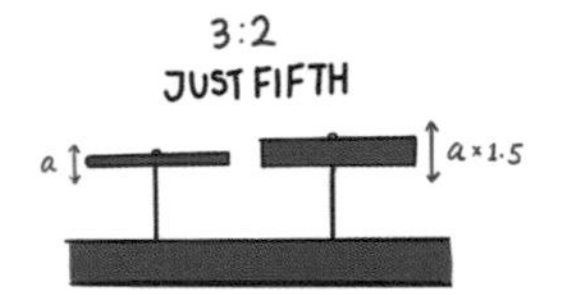

Understanding numbers and ratios facilitated developing music theory, as numbers underlie harmony and music. For the Pythagoreans this further reinforced their conviction that "all is number". Not just any number. Whole number – and the ratios between whole numbers, of course.

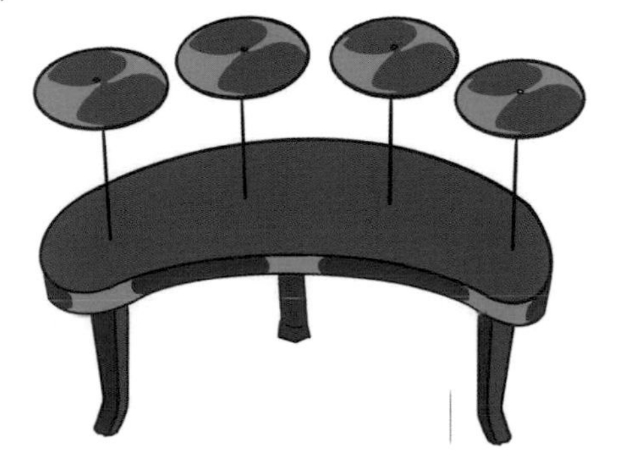

HIPPASUS'S MATHEMATICS

Wholeness of a number was of great significance. This was similar to Democritus's idea of the "atom" in the physical world, the building block of everything – something that cannot be further cut. Likewise, a whole number dictated how everything could be described. In this peculiar "number atomism", ratios between integers were also allowed – the measures would be *commensurable* if the ratio that described their relationship consisted of integer parts. In geometric terms, any two lengths would be commensurable if there was a third length than would fit in both lengths an integer number of times.

The matter of (in)commensurability may have started emerging through Zeno's paradoxes that essentially rendered all movement impossible. In the famous paradox, Achilles being a superhero allows the tortoise to start the race first (a superhero that competes with a tortoise nonetheless). Each section of the racetrack between Achilles and the tortoise could have been halved over and over again. Achilles never reaches the tortoise, which means Achilles is essentially stuck unless he starts concurrently with the tortoise. Of course, this contradicts our experience, but this thought experiment gives rise to a consideration of relationship between measurements, and how long one can keep dividing a given quantity.

Talking of division, think of long division: how long would you carry on adding zeros and calculating more decimal places before you give up? When will you be prepared to admit that the digits must be going on forever? Especially if the digits are not displaying any visible patterns. If you keep getting threes for example, it feels somehow less daunting, maybe even slightly reassuring in that the pattern indicates some expectation that the answer will be a nice enough number. But what if there is no pattern visible whatsoever? Would you give up then? Well, actually, you should. Not that it would kill you or anything . . .

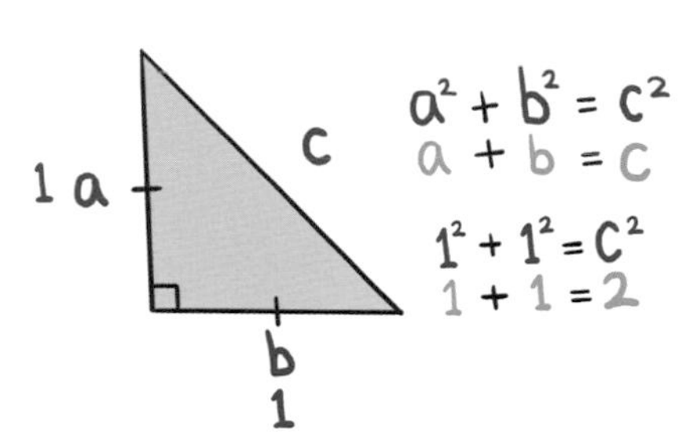

Or so Hippasus thought. It is unlikely that Hippasus anticipated coming across a mathematical finding that would defy the foundation of the Pythagorean belief system, even if he favoured fire more than number. Amongst other geometrical problems, he studied the properties of a right-angled triangle that was also isosceles.

Let us just assume that the two perpendicular sides were one unit each, whichever unit was used at the time (that's not important). Using Pythagoras' theorem, what would the length of the hypotenuse be? Well, $\sqrt{(1^2 + 1^2)}=\sqrt{2}$. Now try to write $\sqrt{2}$ as a ratio of whole numbers (or a fraction) if you can: you will be failing for ever. The square root of two is an irrational number, meaning that there are no two whole numbers that when divided by one another will result in $\sqrt{2}$. So, we know the size of $\sqrt{2}$, we know how to draw it, and where it lies on the number line, but we also know that it cannot

$$\sqrt{2} = \frac{?}{?}$$

$$? : ?$$

be written as a fraction. This was revolutionary, far more than any previous discoveries; this one was changing the meaning of number itself.

Do not think this only happened once in history. People had difficulty accepting zero (how can nothingness be a number?), negative numbers and imaginary numbers ("imaginary" was in fact the term they used to mock the mathematicians who first perceived numbers that were roots of negative numbers, we will explore a bit more on this later in this book).

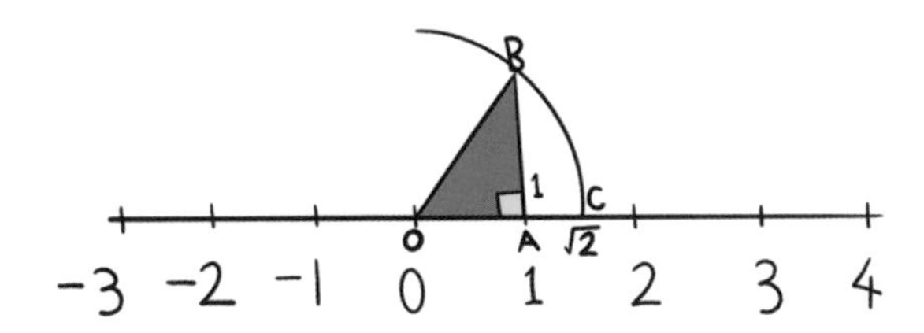

Here is the proof that the square root of two is irrational. The method utilised here is called proof by contradiction: that is assume something is true, and using it arrive to a contradiction, hence demonstrating that the original assumption is incorrect. If we assume that $\sqrt{2} = \frac{n}{m}$ where n and m are co-primes (their HCF[3] is 1), it means that $2m^2 = n^2$, which

3 Highest Common Factor

means n^2 is even (it is equal to a multiple of two). And if n^2 is even then n must also be even. So n^2 is in fact a multiple of 4. That, in turn makes m even too. So, n and m are not co-primes. And this contradiction allows the initial assumption to collapse: the square root of two cannot be written as a fraction.

Proof by contradiction that the square root of 2 is irrational:

Let $\sqrt{2}$ be rational and that
$\sqrt{2} = \frac{n}{m}$*, in its lowest terms*
$m^2 = n$
square both sides to get
$2m^2 = n^2$
So n must be even
(a square number can only be even if the original number is even)
and n^2 must be divisible by 4
So m^2 must also be even
which implies m is also even
So $\frac{n}{m}$ is not reduced
(at least 2 as a common factor)
Hence $\sqrt{2}$ cannot be written as a fraction

A plausible question emerging from Hippasus's finding is why didn't this discovery cause an earlier abandonment of the "all is number" maxim a bit sooner? Maybe it was a matter of pride and what would be compromised by admitting one is wrong. This intolerance towards Hippasus's finding is the most probable cause of what came next.

THE DEATH OF HIPPASUS

Hippasus's discovery of irrational numbers caused great distress to the brotherhood. The teachings on anything that had to do with numbers would have to be re-examined. But why re-examine something that seemed to have worked and convinced people of its validity? This would be just too upsetting for the members of this community.

To top it off, Hippasus, despite having sworn secrecy and devotion to Pythagoreanism, allegedly divulged his finding outside the group, which compromised the group's reputation and reliability. Did he really deserve what happened to him? It was less than fine day to cruise the Aegean Sea, and those days are rare in the Mediterranean climates. Hippasus found himself drowning after being thrown overboard – an immolation to Poseidon, or an act to appease the elders of the group? And all for a mathematical truth which could not have been hidden for very long, anyway!

AND NOW A BIT OF STATISTICS

Percentage of deaths by drowning: 7% of all injury-related deaths

---

Percentage of deaths from revealing a mathematical truth: negligible

---

Probability of deaths for publicising without permission: unlikely, however many such disputes in the past have resulted in bitterness and lost jobs.

YOU SHOULD NOT BE DISTURBING MY CIRCLES!

ARCHIMEDES'S LIFE

Archimedes lived in the 3rd century BC in the wider area around the Mediterranean, including his hometown Syracuse, and Alexandria. He was exceptionally prolific: a mathematician, physicist, engineer, inventor, and astronomer. His understanding of mechanics was so sharp that he was comfortable boasting that given a long enough lever and a suitable place to stand, he could move the earth.

Amongst his inventions was the Archimedean screw pump which allowed for water to be transferred from a lower place to a higher. The other way around had always deemed to be fairly straightforward.

He also invented a geometric puzzle, similar to the Chinese tangrams, called Ostomachion. It was made of ivory (not illegal at the time). The word possibly comes from "osto", meaning bone, and "mache" meaning battle. However, through transliterations, there was a possibly unfortunate connection with the stomach (as in the body part). Ostomachion was a square cut up in 14 pieces that can be rearranged to form that same square in 536 distinct ways. The pieces can also be rearranged to form all sorts of plants, animals, and more.

Archimedes is perhaps most famous for the principle that bears his name. The story goes as follows. The then King of Syracuse, Hieron the Second, enquired whether the crown he had recently commissioned was indeed pure gold and whether it contained the full quantity that was handed to the goldsmith to manufacture. It turned out that it was not a simple question: Archimedes contemplated on this problem persistently and he only stumbled across a solution when in his bath. The excitement was too overwhelming and Archimedes could not contain himself. Naked as he was, he started running around the neighbourhood shouting "Eureka". Incidentally

in Greek it is pronounced "Evrika". The solution relied upon identifying the density of the crown: dividing its volume by its mass. By immersing the crown in water and measuring how much

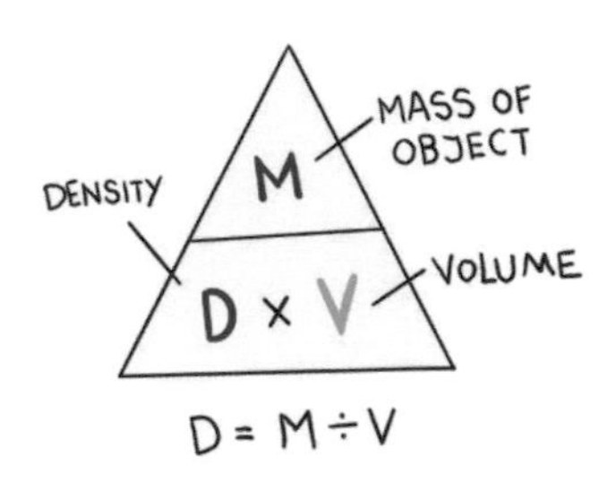

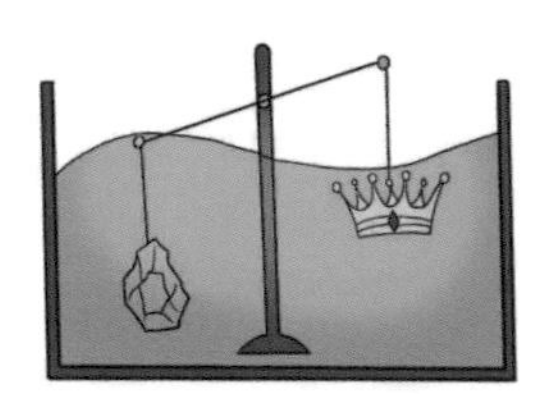

water has been displaced, one can find the volume of an irregular shape such as a crown. The story has it that the goldsmith did cheat. But the goldsmith's death is not on our list.

Amongst his other famous inventions are some that were used at the time when Syracuse was under attack by the Roman army. By using speculums (metal mirrors) held by defending soldiers in appropriate positions, the sun rays were focused on specific points and enough accumulated heat would start a fire. These stories have been disputed by historians (as with some stories about the previous two mathematicians we have met), but there is sufficient evidence to

suggest this was wholly possible. Archimedes also created catapults designed to throw huge stones sinking the ships and hence keeping the enemy away for longer. Syracuse eventually fell to the Romans despite defending using Archimedes's brilliant machines. But you might be glad to know that there are suggestions that he considered his inventions more as a hobby or applied geometry of sorts, while valuing his mathematical studies above all else.

ARCHIMEDES'S MATHEMATICS

Archimedes' output was copious in all areas of his endeavour. Even as recently as 1906 an old manuscript surfaced; it was a palimpsest (a parchment used repeatedly after previous writing had been erased) with prayers over it. It took decades and great technological advancements for the earlier layers to be revealed and read. It seems like the methods Archimedes described in that manuscript were very advanced. Had they been available to the mathematicians of the 17th century, they could have accelerated the development of calculus.

Here we will be discussing his work on finding an impressive approximation of π (Pi[4]), the ratio of the circumference of a circle to its diameter. This method resembles the upper and lower Riemann sums, for those readers familiar with the method from calculus (no such knowledge required here).

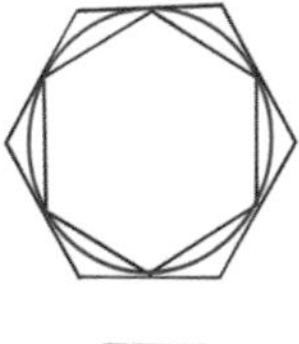

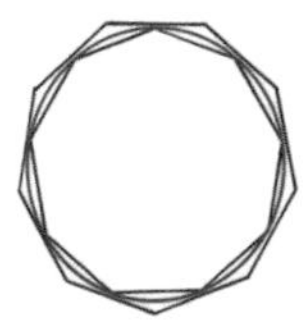

The idea used here is that circles can be considered as regular polygons with an infinite number of sides. Additionally, an inscribed polygon would underestimate the circumference of the circle and a circumscribed polygon would overestimate it. The more sides the two regular polygons have the closer the approximation

4 π in Greek sounds more like "be" rather than "pie". However, being pronounced more like "pie" in English, we had to have a nice fruit jam pie on the page. And guess what, you can find π in that pie just like in any other circle!

will be. So, in Archimedes' time you would have to do it all by hand. How much patience, and accuracy could you summon?

Let's start with a regular hexagon: that's really easy. Draw a circle, choose a random point on its circumference. Put the point of the compass and keeping the opening the same as the radius, mark a point on the circumference. Continue until you have gone all the way around and got back to your original point. By joining all six points you have an inscribed regular polygon.

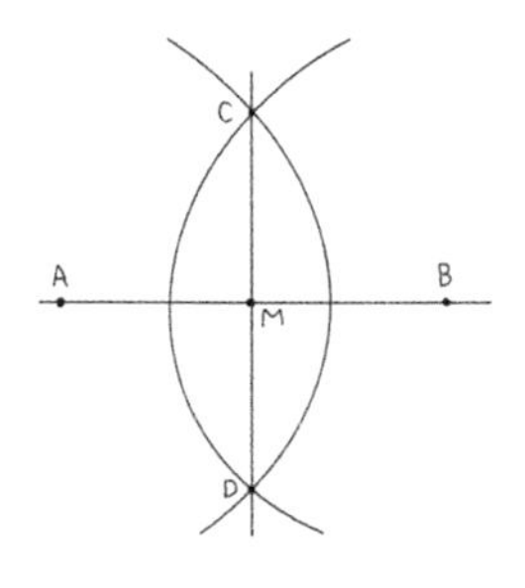

By constructing perpendicular bisectors of all six sides, and connect the now 12 points on the circumference, you have created an inscribed dodecagon. And then, a few perpendicular bisectors later, a regular 24-gon. Then, yes, you guessed it, 48, and then 96.

Archimedes did go all the way to constructing a 96-gon. But this was not the final step! Bear in mind that all these regular polygons were *in*scribed and hence were all underestimating the circumference of the circle. Here's where patience comes in handy: continue to construct circumscribed polygons up to 96-gon too. Ok, I'll explain how, but do not feel obliged to construct them.

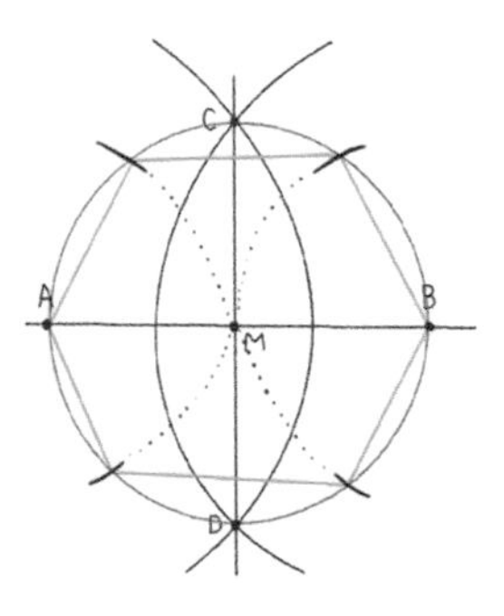

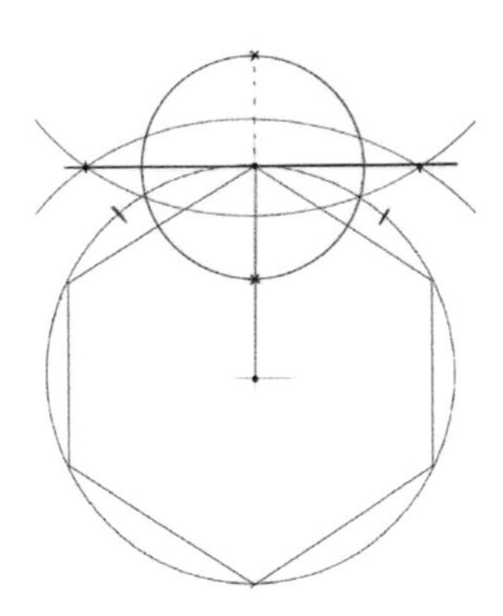

What is needed here is the knowledge on how to construct tangents to specific points on the circle. Starting from the inscribed regular hexagon, take any of its vertices on the circumference. Open the compass on a smaller radius than the existing circle. Using the aforementioned vertex as a centre, draw a new circle. Draw the line that connects the centre of the original circle with the vertex and extent it to intersect the new circle twice. Now draw the perpendicular bisector of the diameter: that's the tangent! Repeat this process six (or 96) times and you're done!

By dividing the perimeters of the inscribed and circumscribed polygons by the diameter of the circle, the approximation of π Archimedes achieved was $3\frac{10}{71} < \pi < 3\frac{1}{7}$, which differs to our current decimal value only in the third decimal place, which is exceptionally impressive. This level of accuracy is rarely useful in practical terms, but some people do enjoy memorising lots and lots of digits of π.

THE DEATH OF ARCHIMEDES

Archimedes did not die peacefully in old age as his stature and fame should have allowed him to. Towards the end of his life Archimedes was residing in Syracuse. And during the Second Punic War and the siege of Syracuse by the Romans, there were specific instructions for Archimedes to be captured but not harmed. The projectile and other mechanisms invented by him to defend the city resulted in a desire for his skills and knowledge to be further utilised by the Romans.

Captain Lucius may have been the soldier sent for the seizure. Archimedes, enjoying his geometric explorations on a sandy beach, (or at home on a wax tablet) was not happy with a soldier persistently asking him to follow him. The compass resembling a threatening weapon is disputable. But Archimedes, demanded the soldier to not disturb his circles (μη μου τους κύκλους τάραττε) and the soldier's patience ran out: he killed him with his sword. What happened to the soldier, we will never know.

AND NOW A BIT OF STATISTICS

Probability of dying by a sword wound: similar to that of general stabbing (see Pythagoras).

Probability of dying during war time: extremely high even if just a civilian (up to 90% of war casualties could be civilians).

Probability of dying because you don't want your study of geometry interrupted: negligible.

WHAT? A WOMAN MATHEMATICIAN? DIE!

HYPATIA'S LIFE

Hypatia is an all-time favourite of mine as she is the first female mathematician to be acknowledged and whose name survived centuries of silencing and oppression. She has become a symbol for equality and individual liberty; nowadays journals and galleries bear her name. There has even been a film made about her life. In case you have watched that you may want to compare the ending of the film with the last section of this chapter. But do not rush to any conclusions!

She lived in the 4th to 5th century AD in Alexandria. Her exact year of birth is disputed, somewhere between 350 and 370, and in the title, naively, you can see the average. Mean or median, but certainly not the mode, as there is no mode in this case. She was a philosopher, a mathematician and astronomer, and most importantly a beloved teacher at the University of Alexandria. Her unique position, unavailable to most females of her time, is owed to her father Theon, who allowed her to be educated as if she were a son. There is little or no information about her mother, however.

Theon was the director of the Museum of Alexandria which resembled a less structured version of a modern university. It might be described as an educational institution, in a multicultural city, attractive to intellectuals from around the world. The significant number of "books" or papyri rolls available made it an even more appealing destination for

the curious and those wanting to learn or progress their careers. Hypatia grew up in these amazing surroundings, being better educated than most, and responding to this education extremely well, finding her own way.

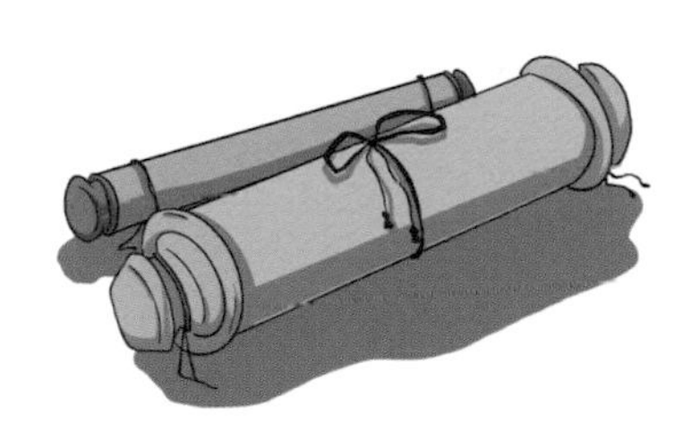

She has been described as a beautiful woman with a well-trained body, and several suitors approached her persistently. Hypatia thought that the best way to make intellectual progress was to abstain any romantic encounters and she chose to remain celibate. There even seems to be an incident where she presented a particularly persistent admirer with a used menstrual pad to repel him.

Hypatia evolved, by all accounts, into a highly gifted teacher who would enjoy great admiration from her students. So alluring and enticing was her teaching that she would regularly have several visitors at her house paying their respects. People would even travel significant distances just to listen to her lectures. She would sometimes wear a tunic, normally worn by male professors and teach around the city, accumulating crowds who found her approach and teachings thorough and meaningful.

One of her most dedicated students was Sinesius. He wrote many letters and those formed a significant source of understanding her teaching style, prudence, elegance and why she was essentially one of the most gifted teachers at least of her time. Many of Hypatia's students seemed to go on and have successful careers, and high-ranking positions. So teaching makes a difference, hmm, who would have thought!

The abundance of faithful students, excellent network, and the fact that she was close friends with Orestes, the Roman governor of the city, caused Archbishop Cyril to develop a certain level of jealousy towards her. She was also a Neoplatonist pagan: Neoplatonism was a philosophical system teaching that happiness is achievable in this life and no after life could necessarily be expected. As Archbishop Cyril was a strong, to say the least, defender of his Christian faith this may well have made things worse.

HYPATIA'S MATHEMATICS

Hypatia has not given her name to any theorems and probably no original writings of hers have survived. This is not because she discovered nothing or because her contributions were not important. It is possibly because some gender barriers were insurmountable. She overcame quite a few, but having theorems named after her was not one of them. She reviewed top male mathematicians' work, such as Apollonius's ("Conics"), Diophantus's ("Arithmetica"[5]) and Ptolemy's ("Astronomical Canon"). The commentaries are essentially editing works, where the editor may add information or further explanations to the original text they have been copying.

This editing work is more important than it perhaps sounds. It helped older writings to survive. Many important works would have been lost if it were not for the work of commentators and editors. Hypatia was an expert in what she edited and taught. She developed certain mastery in those fields and possibly added sections that challenged her students, helping them to develop their mastery further too.

The following pair of simultaneous equations were probably added to Diophantus's work by Hypatia:

$x - y = a, x^2 - y^2 = (x - y) + b$, where a and b are known values.

Diophantus was the first mathematician to consider fractions as numbers and identified equations with no solutions or infinite solutions. Using algebraic tools and more abstract language than the one available

5 It is in a margin of Diophantus's "Arithmetica" that Fermat wrote his famous quote about having a proof for his theorem, but that it does not fit in the margin. A proof was eventually given by Andrew Wiles centuries later, which was tens of pages long.

to Hypatia at the time, we solve this pair of a linear and a non-linear equation with two unknowns using substitution. This method is pretty basic, it only takes a bit of practice.

$$x = a + y$$

Substitute this linear expression into the quadratic equation to eliminate *x* and solve for *y*. Then come back to this to find *x*.

$$(a + y)^2 - y^2 = (a - y - y) + b$$

$$a^2 + 2ay + y^2 - y^2 = a + b$$

$$2ay = a + b - a^2$$

$$y = \frac{a + b - a^2}{2a}, \textit{and}$$

$$x = a + \frac{a + b - a^2}{2a}, \textit{or}$$

$$x = \frac{a^2 + a + b}{2a}$$

There is evidence suggesting Hypatia was also an expert user of the astrolabe, so much so that she introduced new features to it. An astrolabe is an instrument that allows one to tell their exact position and direction based on the positions of the celestial bodies, like an ancient GPS. It consists of a round thick plate with graded circumference and further plates with different information fixed over it around a central axis, including a 2-dimensional representation of the stars in the night sky. On its back, there is a straight-line segment with dioptres that the observer would align with their celestial foci. With appropriate rotations and alignments with celestial bodies, one can predict the rise and setting of a star. An astrolabe could allow one to make predictions on lunar and solar eclipses. And that is where things may have gone really wrong for Hypatia.

THE DEATH OF HYPATIA

The Christian zealot groups were not thrilled with this audacious woman talking to devoted students. Her being able to predict such major astronomical events as eclipses was misinterpreted as her actually causing the celestial bodies to disappear temporarily (just for fun). That led to scaremongering the possibly malleable crowds that she was a witch. Spreading this rumour must have not been too challenging as the popular opinion was that this woman did not know her place. Additionally, her pagan views had put her on the spot already. The evidence that Archbishop Cyril coordinated the rise against Hypatia is precarious, but possibly so were his efforts in preventing it. Either way, his jealousy of her skills and knowledge are fairly well evidenced.

When the power struggle between Cyril and Orestes peaked, and Orestes almost lost his life to Cyril's supporters, he avenged the attack with arresting their leader and torturing him to death. Despite Hypatia allegedly advising Orestes to reconcile, rumours broke that she had instead cast a spell on him and made him only listen to her, making her accomplice to the torturing and what followed.

Hypatia appears to have been removed from her carriage on her way home by a zealot group led by Peter the Reader. She was dragged into a church called the Caesarium. She was stripped naked and using sharp ostraca the flesh was torn off her bones. Her dismembering was followed by the remains of her body to be burnt. Her horrific end signified the end of the Hellenistic era, as after her death, the intellectuals who led their lives and careers in the famous city fled Alexandria. In the years to follow, scientific advancements started shifting towards the Hindu-Arabic world.

AND NOW A BIT OF STATISTICS

Probability in history of a female being dubbed a witch and killed: high; tens of thousands were tortured and killed from antiquity until today.

Probability of torture with ostraca: negligible; death by torture is however significant.

Probability of being murdered by religion fanatics: extremely high, many religious groups have caused high volumes of bloodshed over the centuries.

A BIT OF GAMBLING KILLED NO-ONE, EVER

GEROLAMO CARDANO
(1501-1576)

CARDANO'S LIFE

This Italian polymath was born right at the beginning of the 16th century. After completing his quadrivium, the standard university "degree" of the time, consisting of arithmetic, geometry, astronomy and music,[6] he continued with medicine at doctoral level. Anything following the quadrivium at the time would have to be at doctoral level and Cardano was exceptionally gifted. However, he displayed difficult character traits early on. This made his interactions with his surroundings often challenging. His rather obnoxious character meant that his reputation was not the best, but he did not seem to pay much attention to this. He remained focused on his reading, studies and prolific publications. He seemed to have an obsession with making history, and he expressed that openly at times.

Despite his exceptional academic skills, he had a hard time acquiring a suitable position at a university because of his illegitimacy.[7]. Furthermore he had upset a number of members of the Milan medical college because of his published disapproval of their practices. However, the fact that his treatment approach to a long-suffering

6 These four were considered branches of mathematics since antiquity. Music fans don't resist, it's true: notes in bars are fractions added up correctly to the same total over and over again.

7 This means that his parents were not married when he was born, and this was a big problem 500 years ago.

Augustinian prior was the only one that seemed to offer the patient any relief gave him a significant advantage as it improved his reputation. This, alongside his persistence, and the public criticism regarding his marginalisation resulted in him eventually getting such a position. Albeit disliked by many, he was gaining status and authority in medical circles.

He continued working with a fierce appetite for success and fame, including having some work published in Lyon, and thus his reputation began to break out of Italian frontiers. Following this, he was summoned to help with the health of the Archbishop of St Andrew's in Edinburgh. John Hamilton had been unwell for several weeks and asked for Cardano specifically. Cardano travelled for weeks and arrived to find an Archbishop on the brink of death. After applying several changes in the patient's daily routine, and persevering caring for him for almost six weeks, he left him well enough to live for several years more. And Cardano made a handsome sum after this successful treatment.

His explorations continued, wide ranging, ambitious, and to us perhaps outrageous. He expanded on previous work completed by a colleague of his on metoposcopy (study of the forehead) He divided it into seven zones and wrote how each planet affects one's character, based on those zones. His love for horoscopes was undeniable; but it was a rather generalised belief that horoscopes have an impact on one's health and as part of devising treatment plans there was an extensive use of astrology. This was common among scientists at the time. More innovative was his concern with finding ways to communicate with the deaf and teaching the blind to read, indicating motivation to contribute in ways that would improve the patient's quality of life markedly.

CARDANO'S MATHEMATICS

His interest in mathematics was rather deep, stemming from lessons with his father and further independent studies outside the quadrivium. Amongst others, he wrote an "abbaco" book on basic arithmetic principles that could be used by teachers and merchants. His abbaco book, "Practice Arithmetice" was one amongst many others but was approaching concepts more formally or at a higher level than usual. It was even written in Latin and was perceived as rather ostentatious; Cardano was never an abbaco teacher or a merchant so his motivation to write this book is unclear; however, it must have influenced the subsequent teaching of arithmetic at university level.

Cardano's encounter with Tartaglia is now one of the most infamous stories of mathematical disputes. Tartaglia was a skilled mathematician, however his social status was nowhere near Cardano's elite medical position. Cardano convinced Tartaglia to share some of his higher degree equation solutions with him. The means he used were not negligible; he swore in rather exaggerated terms that he would never reveal those solutions and would even store them in cipher. Tartaglia, possibly hoping that Cardano would introduce him to key people to help his social ascent, mistakenly trusted him. But this

was a man who wrote a book on probability – with a special section on how to *cheat* when gambling – and was, as you might expect, betrayed. Tartaglia's solution appeared in Cardano's "Ars Magna", his lengthiest mathematical treatise. Tartaglia tried to prove his authorship and challenged Cardano to a debate. Cardano sent his student Ferrari, who won over Tartaglia rather easily, leaving Tartaglia embarrassed and devastated.

Cardano was a passionate gambler, who admitted that he was not just gambling now and then but actually every day. He was so skilled that he would make his living out of gambling at times. He first touches upon issues of chance and winning in "Practice Arithmetice", where referring to what later became known as the problem of points – how to share a bet if the players must end their game early – Cardano insightfully found that it should not be based on what they have already won but rather on the chances that they would have to be the overall winner had they continued. Despite the insightfulness in the principle of the correct solution, he did not compute the final ratio correctly.

His experiential wisdom on the practicalities of probability can be found in his probability book "Liber de Ludo Aleae" (or the book of games of

chance). Cardano argues there that gain from gambling is legitimate as long as the opponents are aware of the risk and skilled enough to play. At a time when the Church did not approve of gambling Cardano dedicated his efforts to show that there is nothing wrong with gambling. To make things more challenging, Cardano included a section on ways to cheat such as advice on rounding off corners of dice, making faces flatter, or even creating dice with rectangular faces (though interestingly enough he did not discuss weighted dice). He – perhaps – disguised this section as things to be aware of so as not to not become the victim of cheating. Hence, we cannot flatly conclude that he was encouraging cheating.

In his probability book Cardano provided the first formal definition of probability as the proportion of favourable outcomes out of a given circuit (sample space). This is nowadays considered rather obvious but stating it for the first time gives Cardano another little bit of legacy, though not widely acknowledged. His distinction between luck and chance is less clear-cut than in subsequent rigidly mathematical treatises of the concept by Pascal and Fermat, hence Cardano was not regarded as highly when it came to the history of probability as his two successors.

He explored problems such as the probability of getting an even number in three successive throws of the dice. While instances of getting an even or an odd number in one throw are equally likely, the same rule does not stand when one requires three successive throws.

Cardano was using a ratio of favourable to remaining outcomes, which in the case of even to odd would be 1:1. However "cubing" the ratio would still give 1:1 which is of course an incorrect result. In modern notation the probability of getting three evens would be $(\frac{1}{2})^3 = \frac{1}{8}$, which is obviously far less likely than Cardano's original calculation.

Even if one looks at the appearance of a specific face *at least once* in three successive throws, that would give a probability of[8]

$$ynn + nyn + nny + yyn + yny + nyy + yyy = \frac{91}{216}$$

$$\frac{1}{6}\times\frac{5}{6}\times\frac{5}{6}\times\frac{5}{6}\times\frac{1}{6}\times\frac{5}{6}\times\frac{5}{6}\times\frac{5}{6}\times\frac{1}{6}\times\frac{1}{6}\times\frac{1}{6}\times\frac{5}{6}\times\frac{1}{6}\times\frac{5}{6}\times$$
$$\frac{1}{6}\times\frac{5}{6}\times\frac{1}{6}\times\frac{1}{6}\times\frac{1}{6}\times\frac{1}{6}\times\frac{1}{6}=\frac{91}{216}\approx 0.42$$

So there still can be no discussion of equally likely events. It has been reported that a confusion between having three winning options in a single throw or getting a single face in three separate throws has produced some erroneous results in his book.

8 y=yes, n=no

THE DEATH OF CARDANO

Cardano continued to expand his knowledge in several areas and acquiring enemies in even more. Additionally, he became increasingly discontented as his two sons suffered great misfortunes. One, through a bad marriage that Cardano had strongly opposed to, ended up in prison and was subsequently executed. The other son, through not reading his father's book and gambling most of the inherited fortune away, brought his family further dismay.

Despite the desperation caused by his family situation, or even further fuelled by it, Cardano, with no subtlety in his manners, continued to cause fury. He wrote a book against the Church, identified the horoscope of Jesus Christ, and was thrown in jail for heresy. When released, he wrote his autobiography; his faith in astrology led him to (not very arrogantly) predict the date of his own death. And he did die on the day he predicted. However, he probably committed suicide (possibly by drinking poison) to avoid the embarrassment of admitting he was wrong.

AND NOW A BIT OF STATISTICS

Probability of dying from suicide: double to that of homicide

---

Probability of dying so that people do not find out you were wrong: negligible (hopefully)

Probability of dying after drinking a glass of poison: almost 100%

A VERY RICH WAY TO DIE

TYCHO'S LIFE

This Danish[9] mathematician and astronomer lived in the 16th Century, making it just to the turn of the 17th. At the age of two, he was kidnapped by his very wealthy childless uncle. His parents did not search for him, so this might have been arranged, but it is unclear why. His parents had many more children after him. His uncle raised Tycho as his own child and Tycho inherited nobility from him; nobility is amongst the reasons why Tycho is best known by his first name, and we shall use this for this chapter. Tycho was provided with access to fine education from an early age on. Whilst Tycho's initial studies focused on law, witnessing a solar eclipse as a young teenager had a lasting impression on him. It drew him into mathematics and astronomy so much so that he thought mathematical formulae are important enough to be worth fighting for.

A mathematical disagreement emerged when the Danish nobleman Manderup Parsbjer gave either a different version of a formula, or a

9 Tycho's birthplace became Swedish shortly after his death and some may like to also think of him a Swede.

different version of the position of the celestial bodies. Or perhaps there was an insult about who is a better mathematician. What is certain is that there was a duel and in the fight Tycho lost the bridge of his nose and spent the rest of his life wearing a prosthetic one. There are anecdotal suggestions that he had a special nose made from expensive metals to wear on special occasions.

Tycho was in possession of around 1% of the wealth of Denmark and the Danish government allowed him to use a private island, Hven, as his observatory. This setting became one of the most prestigious settings for astronomical research of its time. The fifty families of peasants already living on the island became subject to the new regulations. They had to look after produce, cleaning and other aspects of daily life. Tycho also had highly trained helpers to assist with taking and recording measurements. Amongst the people who were assisting Tycho in his explorations of the heavens was his sister Sophia; it was still rare for women to be involved in studies of any kind and this was a unique opportunity for her.

Tycho was quite authoritative and specific on how he wanted things to run on Hven. He even had a new castle built for him, called Uraniborg,

named after the muse of astronomy, Urania. Apart from extensive observatories and several amenities, including entertainment, a private jester and a pet elk,[10] he also had an alchemy laboratory and a prison for guests that were not as collaborative as he demanded. Near the end of his stay on Hven, he kept a tenant and his entire family in chains against the decision of Denmark's High Court of Justice. His authoritarian attitudes and the vast Royal money he managed led to him becoming gradually alienated from those who surrounded him. While the exact cause for him leaving Hven after 21 years is unclear, the ascent of a new king to the throne and the loss of support from the Crown may have been part of why Tycho abandoned his magnificent mini-kingdom. After a short time in Germany he went to Prague, now in the Czech Republic, where he spent the rest of his life.

10 The elk accompanied Tycho to a dinner at another nobleman's premises, drank a lot of beer, fell off the stairs and died.

TYCHO'S MATHEMATICS

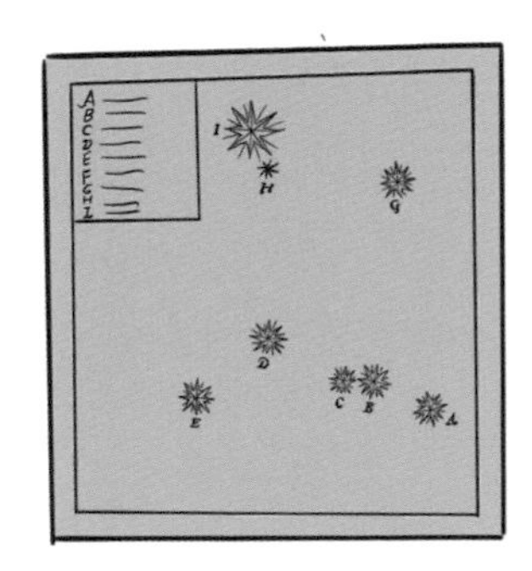

The prediction and occurrence of the solar eclipse of 1560 must have compelled young Tycho's curiosity. His mathematics professor provided him with Ptolemy's *Almagest*, the only printed mathematical and astronomical treatise available at the time. A subsequent erroneous prediction of a Saturn-Jupiter conjunction seemed to have convinced him that accumulating accurate measurements of the heavenly bodies was worthwhile. The final straw that seemed to have a decisive effect on his choice of career was the discovery of a *nova stella*, nowadays called supernova. Despite the high accuracy of his measurements, this celestial entity seemed to maintain an unchanged parallax, indicating how far it must be. Tycho was more dedicated than ever, and this led to his legacy thereafter: a massively impressive collection of unprecedentedly accurate measurements, alongside the development of suitable instruments to allow him to do so.

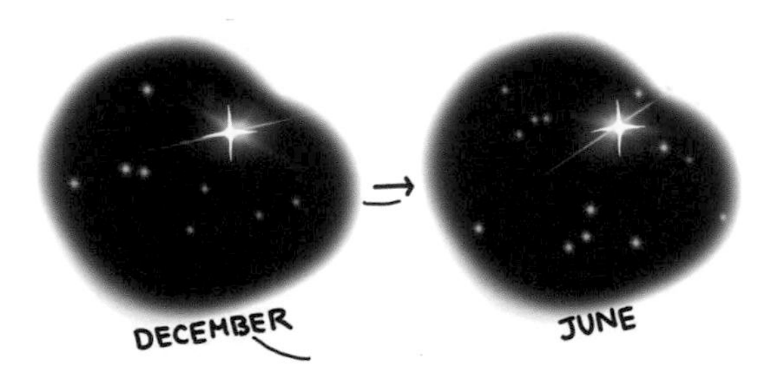

But what is a parallax? Stellar or trigonometric parallax is what is perceived as movement of a star in relation to its background. The closer a star is to us, the larger its perceived parallax is. Therefore, if a star is observed with a certain background in December, and then again six months later in June, as the Earth has moved in its orbit of the Sun, the background stars will have changed. Half the angle formed between the star and the two positions of the earth is the parallax. Hence the further a star is the smaller the angle formed. For stars extremely distant (thousands of light years[11] away), the angle formed will be tiny and the background would barely change. Parallax relies on good old geometry and trigonometry; knowledge of the radius of earth's orbit and the measured size of the angle will allow for the distance of the stars to be calculated. Tycho used this method to show that stars and comets were positioned far beyond the moon.

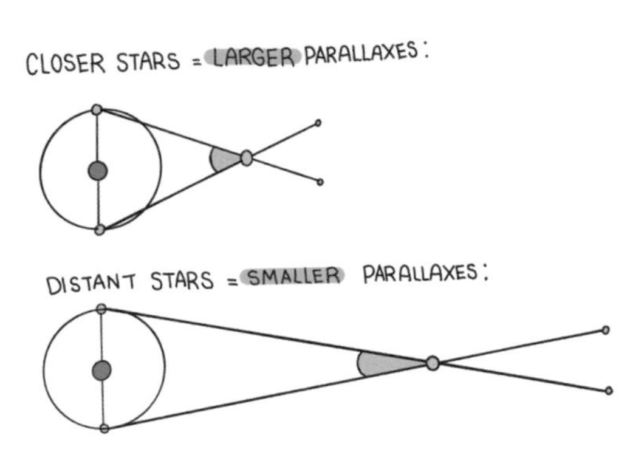

Tycho's wealth allowed him to expand his astronomical studies significantly. Over his time in Hven, he constructed around 30 instruments and significantly improved the sextant allowing him to perform highly accurate measurements before the telescope was invented. The quadrant and the secant were exceptionally important. He kept detailed records and measurements of several hundreds of stars. The meticulousness and accuracy of his records did not seem to result in any significant new theories. However, he did strongly advocate for a move from

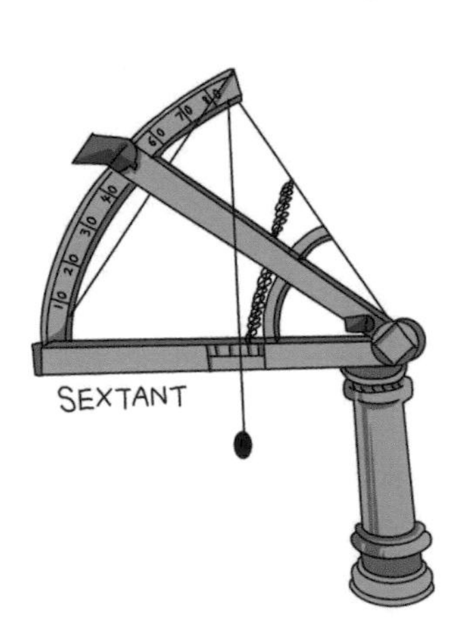

11 One light year is a unit measuring distance and not time; it is the distance light covers travelling one year. That is approximately six trillion miles, or really *really* far.

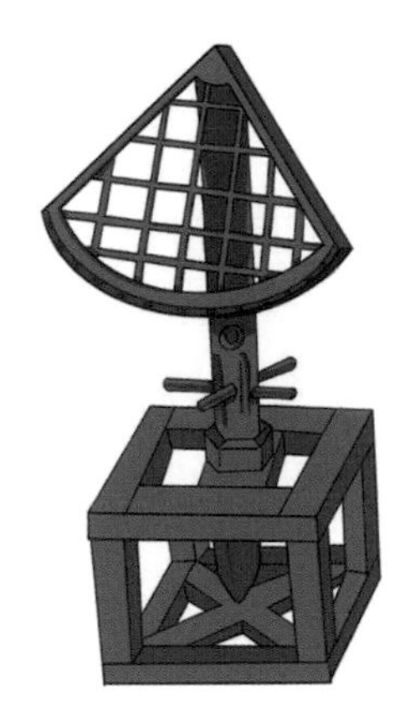

geocentric to geo-heliocentric cosmology. In geo-heliocentric system, the planets move around the sun, but the sun moves around the earth. The geo-heliocentric model is still called Tychonic today. At its time it was popular enough that Ursus, an astronomer from Prague, was accused of attempting to plagiarise his model. Access to Tycho's masses of data, including data on Mars, allowed his then assistant Johannes Kepler to suggest that the planetary movement is elliptical and not circular, and subsequently state his famous three laws of planetary movement.

THE DEATH OF TYCHO

Whilst he was the personal astrologer for the King of Denmark, he was suspected of being the Queen of Denmark's special companion and either the King himself, or the prince ascending to the throne may have wanted him dead. The high concentration of mercury found on his moustache and his hair remained inexplicable, so much so, that his remains have been re-examined long after his death. The re-examination of his remains on two occasions in 1901 and 2010 took place in order to establish if he was murdered or not. Scientists believe that there is not enough evidence to suspect murder as even the mercury found could be justified by his alchemistic experiments.

Whilst a guest at a banquet in Prague, he was served drinks that may have been compromised. What was worse, he refrained from relieving himself for hours in order to avoid insulting his hosts. Upon his return home, he found himself unable to urinate except in the smallest of quantities, and after eleven days of excruciating agony he finally died, possibly after his bladder burst.

AND NOW A BIT OF STATISTICS

Death from inability to urinate: Not that rare: within one year one in four men with urinary retention die because of this condition.

Death from spiked drinks: unlikely, but can lead to a comatose state.

Death from too much mercury on your moustache: well, that would require a prosthetic metal nose that could react with the chemicals in your alchemy laboratory, so let's say, negligible.

DEATH BY TIME CALCULATION

ABRAHAM
DE MOIVRE
(1667-1754)

DE MOIVRE'S LIFE

This French mathematician was born in Vitry-le-Francois. His family, with reasonable but not exceptional wealth, valued education very highly and prioritised accordingly. Despite being a Protestant, his first schooling was at a Catholic school. He studied mathematics mostly in his own time rather than as part of his initial formal schooling. However, he had to change schools because his family had to move around, not unlike these days.

Huguenots in France[12] suffered significantly; they were marginalised or even prosecuted as this religion was illegal since the King revoked all their rights in 1685. This led to many Huguenots deciding

to flee across the channel, and so did De Moivre. The most likely year of De Moivre's arrival in England was 1687 although some sources suggest he was imprisoned briefly in France, purely for his religious beliefs. He was now in a land where he could express and

12 Huguenots were a particular group of French Protestants.

practise his religion freely, but nonetheless a foreigner, which of course carried a whole different set of difficulties.

In London, he met a few influential people, most notably Isaac Newton[13] and Edmond Halley[14]. De Moivre enthusiastically bought his own copy of the newly published "Principia Mathematica" and studied it thoroughly. It is known that due to lack of time – he was working as a private mathematics tutor around London – he would cut out a few pages of the bulky book at a time and study those while travelling between destinations. Newton himself was impressed on how well De Moivre comprehended the content of his book and was openly complimentary about it. Newton had such respect for his ideas that he would invite him to his house, and they would discuss philosophical matters.

Upon inspection of De Moivre's work and mathematical contributions, Halley found they were of such high quality and sophisticated enough for

13 1643-1727, English mathematician, physicist, astronomer, author of ""Principia Mathematica"" etc etc. A most unnecessary footnote indeed! But let's add one of the best quotes in history to make the footnote worthwhile: as Newton said "If I have seen further than others, it is by standing upon the shoulders of giants".

14 1656-1742, Halley's comet was named after this English astronomer.

them to be presented to the Royal Society. Shortly after, De Moivre was voted a Fellow of the Royal Society. However, he never managed to acquire a teaching position at a university, and he remained a (highly esteemed) mathematics tutor around London for the rest if his life, either at houses or coffee shops.

De Moivre was a regular at Slaughter's coffee house on St Martin's Lane in London, where many Frenchmen (yes, *men*; as this was still the 18th century!), and especially Huguenots, met regularly. There is even a record of him having requested and received post at this place, indicating how frequent his visits must have been. Slaughter's coffee house was also the meeting place for chess players and artists. There were newspapers available and there were often interesting conversations going on amongst the patrons.

He kept regular correspondence with mathematicians, most notably perhaps with Johann Bernoulli,[15] whom he kept up to date with any news from the scientific scene in London. It is suggested that he had influenced Bernoulli's being voted into

15 1667-1748, Swiss mathematician, member of a very large family of mathematicians.

the Royal Society as a fellow. He was also involved with the infamous Newton versus Leibniz dispute on who invented calculus.[16] As Bernoulli supported Leibniz, unlike De Moivre, their correspondence ended around this time.

16 Newton and Leibniz presented their results on calculus, with Leibniz publishing first but Newton claiming that he plagiarised as the two had been in regular correspondence regarding mathematical matters previously. At the time it seemed crucial that the discovery is attributed to only one of them, and Newton won that battle, and the Royal Society made an official declaration attributing the invention of calculus to Newton in 1715. This led to the complete discrediting of Leibniz in the eyes of his contemporaries, and he died the next year. Nowadays research has shown that the two approaches were different enough to safely suggest that the discoveries were indeed independent.

DE MOIVRE'S MATHEMATICS

De Moivre contributed to several areas of mathematics resulting in theorems carry his name today. In "De Mensura Sortis" from 1711, he offers an accurate definition of probability using terms similar to favourable and unfavourable outcomes. The success to failure ratio would be the same as the ratio between the possible ways achieving the desirable outcome to the number of ways to fail.

De Moivre's "The Doctrine of Chances: Or a Method of Calculating the Probability of Events in Play", which was published in 1718, contained results on games of chance and was particularly well received by gamblers of his time. De Moivre seems to have been offering advice on games of chance orally as well; whether he was paid or not for this advice, is debatable.

His book describes calculations of probability for independent events for the first time. Independent events (often confused with mutually

exclusive[17] events) are events that do not affect one another. Say, whether or not Cardano committed suicide does not affect whether or not De Moivre also did. Therefore, these two can be described as independent events. However, they are not mutually exclusive; one or both or neither could have happened just as well. The formula for independent events is as follows:

$$P(A \text{ and } B) = P(A) \times P(B)$$

Independent Events (A, B) — Probability of both occuring

If the probability of *both* events happening is the same as the product of the probability of each separate event, then the events have no effect on one another and vice versa. In fact, adding "vice versa" is not trivial in mathematics. It means that a condition is necessary and sufficient for something to be true. So, if you know the events are independent, you can use this formula. And if you know the formula turns out to work after you substitute given values in, it means the events are independent. You can also say two events are independent *if and only if* the probability of their intersection[18] is the product of their probabilities. *If and only if* can be written shorthand as *iff.*

De Moivre's book also included results ahead of its time. Such an example is the central limit theorem which did not receive enough

17 Cannot happen together – the occurrence of one excludes the occurrence of the other.
18 Intersection is where the events overlap or a fancy way to say "strictly both".

attention until Laplace reworked and expanded the proof for more values several decades later. It now bears both these mathematicians' names. The idea behind this theorem is that calculating the mean of samples from a variety of different distributions will result in a normal distribution of the means. That is the middle values occur more often than any lower or higher values. You may have seen the normal distribution curve (or Bell curve) before. A simple example of the normal distribution curve would be that there are far more people with an average height than there are very tall or very short people.

De Moivre also offered an early solution to the "Knight's Tour" problem, a question on how a chess knight can go over the entire chess board. The Knight's Tour solution de Moivre found has a starting and ending position far apart from one another. Mathematicians have subsequently improved on his solution.

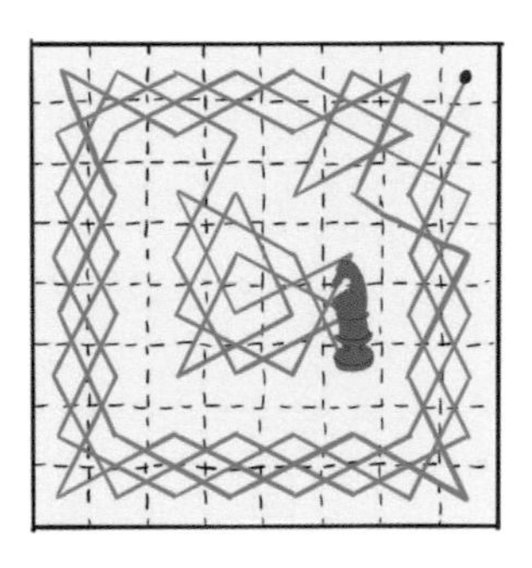

A famous result by De Moivre came from his work on combining trigonometry and analysis; the formula $(\cos x + i \sin x)^n = \cos nx + i \sin nx$ still bears his name. This formula uses trigonometry to find n^{th} roots of complex numbers (real numbers being a subset of complex numbers of course).

THE DEATH OF DE MOIVRE

Unlike common perceptions, De Moivre did not die in poverty. He was earning indeed much less than his contemporary mathematicians with tenure positions at educational institutions. But his will left bequests to his grandnieces totalling around £1600 at a time when the annual salaries were around £60.

De Moivre, very much like Cardano, is remembered for predicting the day of his own death. He approached this slightly differently though: he noticed that the duration of his sleep had started increasing by 15 minutes each day. So, using the arithmetic progression formula, he found the nth term that would equal 24 hours. And indeed, that was the day he never woke up again.

AND NOW A BIT OF STATISTICS

Probability of dying in your sleep: Not as high as one might think; it is actually rather rare unless there are underlying conditions.

Probability of predicting your own death: It seems that feeling close to dying makes it more likely to die, especially the older individuals get, so not negligible.

Probability of dying of old age: Certain, if nothing else has killed you in the meantime of course.

JUST A BIT TOO YOUNG

GALOIS'S LIFE

Evariste Galois was born in Bourg-la-Riene, France. His childhood was peaceful; he grew up with supportive parents who sought to provide him the best possible education. His mother home-schooled him for the first few years. She was of liberal ideology and passed her beliefs on to Galois. Along with his father, a boarding school director and later the mayor of their small town, she instilled in him a deep loathing of tyranny.

His school, College Royal de Louis-le-Grand in Paris, was a trigger for the awakening of his political inclinations because of its overwhelming rigidness. It was, however, to this same school Galois returned to after his premature attempt to enter the Ecole Polytechnique (then, as now, France's leading technical university) failed. Galois had by then published his first mathematics paper at the age of 17: "Proof of a Theorem on Periodic Continued Fractions," in the journal *Annales de matématiques pures et appliqués*. His maths teacher Louis-Paul-Emile Richard, a supportive individual who had a record for encouraging his pupils towards greatness, spotted Galois's exceptional mathematical understanding and directed him further towards suitable reading. Galois published some more articles in the journal *Bulletin des sciences mathématiques, astronomiques, physiques et chimiques* subsequently.

However, his attempt to have a paper accepted by the French Academy of Science was unsuccessful and so was his second attempt to enter the Ecole Polytechnique. The paper was returned with advice by Simeon Denis Poisson[19] that Galois reworked it further, as he tended to omit significant steps and his thinking was very difficult to follow (he didn't show his working – does this ring a bell?). Cauchy[20] refused to read a paper Galois had sent him, and Fourier[21] died (possibly a bigger misfortune for Fourier on this occasion) before he was able to read what Galois had sent him. Any opportunity for recognition kept slipping away from Galois, whose mathematical production

19 Simeon Denis Poisson (1781-1840) French mathematician, engineer, and physicist.
20 Baron Augustin-Louis Cauchy (1789-1857) French mathematician.
21 Jean-Baptiste Joseph Fourier (1768-1830) French mathematician.

needed some outlet. The evidence is consistent that the presentation of the materials was extremely unclear and Galois was unreceptive to feedback asking to improve the papers.

At the same period, during politically turbulent times, his father found that his own name had been forged on malicious epigrams addressed to his wider family. Amidst growing pressure, he committed suicide and that cost Galois dearly.

After he failed to enter Ecole Polytechnique or get his second paper published, he entered what is now known as Ecole Normale, also an elite institution, but with an emphasis on government administration. During Charles X's exile and subsequent rise of Louis Philippe, students were prohibited from joining the opposition. Galois not only joined but his actions were perceived as threatening to the new King and he was arrested. A clever lawyer saved him from a prison sentence but not for long; Galois later attended a protest in the uniform of an outlawed organisation. Eight

months into his incarceration, and during a plague epidemic, he was moved to a care home due to ill health. Things were not meant to improve subsequently.

GALOIS'S MATHEMATICS

Much of Galois's mathematical understanding came from extensive independent studies, some recommended by his teachers, as he would quickly surpass the level of the other students in his cohort. His mathematical engagement was with (then) recent findings in mathematics; Legendre's[22] geometry and Lagrange's[23] work on algebraic equations and analytic functions. Galois's work expanded on Lagrange's findings on algebraic equations. One of his findings had to do with solving polynomial equations using radicals. That is, equations containing a single unknown, say *x*, raised at ascending powers, with rational[24] coefficients. To solve these equations would mean to find a general method containing the four main operations of arithmetic (addition, subtraction, multiplication and division) and appropriate radicals (square root, cube root etc) that would work *every* time.

Whilst solving linear equations is easy and quadratics have a formula simple enough to remember, the general cubic and quartic can be only solved with formulae resembling monstrosities that you have probably never encountered before (but you can see below). But how about quantic? Well, after witnessing the last two formulae, it is probably a relief to know there

$$x = \frac{-b \pm \sqrt{b^2 - 4ac}}{2a}$$

$$x = \sqrt[3]{\left(\frac{-b^2}{27a^3} + \frac{bc}{6a^2} - \frac{d}{2a}\right) + \sqrt{\left(\frac{-b^2}{27a^3} + \frac{bc}{6a^2} - \frac{d}{2a}\right)^2 + \left(\frac{c}{3a} - \frac{b^2}{9a^2}\right)}} + \sqrt[3]{\left(\frac{-b^2}{27a^3} + \frac{bc}{6a^2} - \frac{d}{2a}\right) - \sqrt{\left(\frac{-b^2}{27a^3} + \frac{bc}{6a^2} - \frac{d}{2a}\right)^2 + \left(\frac{c}{3a} - \frac{b^2}{9a^2}\right)}} - \frac{b}{3a}$$

22 Adrien-Marie Legendre (1852-1933) French mathematician.

23 Joseph-Louis Lagrange, or Giuseppe Luigi Lagrange or Lagrangia (1736-1813), Italian (French by naturalisation) mathematician.

24 Rational is a number that can be written as a fraction. This would include all integers too, as they can all be used as numerators over a denominator equal to one.

is no general formula a polynomial equation of degree five or higher, save some special cases. And it was Galois who proved this, using a new mathematical concept he conceived, the group.

The closest mathematical entity to a group you may have seen before is the set. A set is a collection of elements (numbers, animals, books or whatever takes your fancy) that are unique and share a common feature. With some additional properties, a set becomes a group, which is a powerful notion to examine properties of all sorts of cases in mathematics.

A set, combined with an operation[25], will be called a group if it fulfils certain properties. When combining the elements of the set using the operation any resulting element should also part of the original set. Additionally, all the elements should have an inverse, combined with which they will result in the *identity* element, also part of the set. The identity element is one that causes no changes when combined with another element using the given operation and should also belong to the group. An example of an identity element would be the number one when used with multiplication; as it leaves the number it gets multiplied with, the same. Finally, this operation needs to be associative.

Associativity means that if you have to perform the operation on three elements, it will not matter if you did the first and second together and then the third, or the second and third together and then the first.

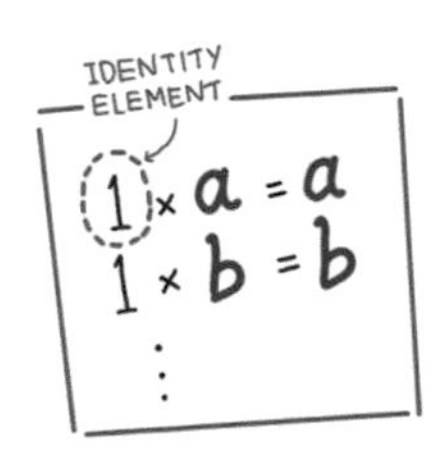

25 The common operations in arithmetic are addition, subtraction, multiplication and division, but for a group an operation could be defined more abstractly.

Depicting it, it would look something like:

ASSOSIATIVITY

$$(a+b)+c = a+(b+c)$$

$$(a * b) * c = a * (b * c).$$

Each polynomial equation has a *Galois group* and the equation is solvable by arithmetic and extraction of radicals if that group is also *solvable* as per Galois' definition. Here things get a bit tricky. Whether a group is solvable or not has to do with the subgroups formed, and how many elements are there in those groups. If the resulting number, called a "composition factor", is a prime number, then the group (and hence the equation) is solvable. The groups are formed with the help of permutations. A permutation shows the ways to arrange different objects when the order matters. If this sounds convoluted, it is! The maths here gets too advanced for now – but maybe understanding more of this could be a future aspiration.

THE DEATH OF GALOIS

Paraphrasing from Murakami,[26] *only the dead stay 20 forever*. Why, after all this hard work and additional hardship, Galois decided to enter the duel that cost him his life is still obscure. The duel happened shortly after him entering the care home, which meant his health was still frail. Was it a romantic entanglement, a setup for a political murder or an awful misunderstanding about something rather insignificant? There's even insinuation that the woman he was involved with was a set-up to lure him into a "duel of honour" that he would almost certainly lose.

Galois, sensing that it would not end well for him, assembled his writings and further annotated as much as possible the night before. One of these papers may have been the one he submitted to the Academy of Sciences. He also wrote two separate letters to his friends. Did the intense mental activity mean he even lowered his chance of survival due to exhaustion?

Galois met Pecheux d'Herbinville, also a political activist, at a secluded place for a duel with guns. Galois did not just lose his nose in the duel. He was shot in the abdomen and was left there unaided until a peasant carried him to the hospital, where he died the day after.

26 Japanese author (1949 -).

His writings were delivered by a friend of his named Auguste Chevalier and eventually deciphered and publicised by a mathematician called Joseph Liouville. His letter ended with the words: "Preserve my memory, since fate has not given me life enough for my country to know my name."[27]

Of course, now the whole world knows his name.

27 In another translation it appears as: "Please remember me since fate did not give me enough of a life to be remembered by my country."

AND NOW A BIT OF STATISTICS

Probability of dying in a duel: Well 50% is that you are the one who gets hit, the rest is down to luck or how well trained your opponent is

---

Probability of dying after being shot: one in four

Probability of dying after being short in the stomach: Less than being shot in the chest but still pretty high, one of the worst gunshot wounds to suffer.

AT THE MENTAL ASYLUM

BLOCH'S LIFE

Bloch was born in Besançon, France, the son of a watchmaker and the oldest of three brothers. He and second youngest Georges were born less than a year apart, so they attended school and then entered the elite Ecole Polytechnique together and were deemed to have comparable mathematical talent.

Soon after the two brothers got accepted in Ecole Polytechnique, the First World War broke out. As they both had military training right after their secondary education, they were conscripted and spent many difficult months with repeated injuries and hospitalisations. Georges lost sight in one of his eyes and was released from duty. André served as a second lieutenant of the artillery at the time. Amidst a fierce bombing by the Germans, he fell from an observation

post and suffered serious injuries. He was given convalescent leave, but for what seems an insufficiently period given the traumatic circumstances he had experienced. At some point André's physical and mental health declined markedly and he was withdrawn from the military and institutionalised. His erratic behaviour led to a decision that he be transferred to Charenton mental asylum in Saint Maurice.

André Bloch (from here on "Bloch") spent over three decades in a mental asylum, following a daily regime religiously. This regime would include sitting at the same seat every day, at the end of a corridor, focused on his studies. He would only leave his seat to have his meals at specified times. There is no record of him seeking to leave the asylum. He was quiet

and collaborative, an ideal patient as described by the medical staff. He did not even want to go outdoors, and he was contented with studying mathematics.

Bloch studied books of advanced mathematics, producing his own notes and conclusions. Furthermore, he kept correspondence with prominent mathematicians of the time to stay up to date. His correspondents included Jacques Hadamard,[28] Gosta Mittag-Leffler,[29] George Polya[30] and Henri Cartan.[31] Bloch's absolute dedication to his studies and regular correspondence put him in a strong position to constantly produce mathematics in a wide range of areas, and to publish articles in reputable journals. Bloch even published at least one paper in collaboration with George Polya, at a time when Polya was unaware of Bloch's living situation and was rather surprised when he found out.

28 Jacques Hadamard (1865-1963) French Mathematician.

29 Gosta Mittag-Leffler (1846-1927) Swedish Mathematician.

30 George Polya (1867-1965) Hungarian Mathematician.

31 Henri Cartan (1904-2008) French Mathematician.

* As you can see in notes 28-31 most of the mathematicians here lived very long lives and did not face particularly peculiar deaths!

BLOCH'S MATHEMATICS

Bloch's work was prolific. His contributions that bear have his name include Bloch's theorem, the Bloch function, and the Bloch constant. It is worthwhile remembering that the closer we get to the present time the more complex and advanced mathematics is becoming – a long way from school mathematics! Let us see the wording of Bloch's theorem:

> Let F be analytic in the unit disc D and F'(0) = 1. Let further BF be the supremum of all numbers $r > 0$ with the following property. There exists a complex number a and a domain $\Omega \subset D$ such that F is injective on Ω and $F(\Omega) = \{z \mid |z - a| < r\}$. The infimum of these numbers BF is positive.[32]

Let us just acknowledge that it is just impossible to explain this work in simple school mathematical terms. Instead, to give a glimpse into what Bloch was doing we will talk a bit about complex numbers, and functions, shedding some light into his work and hopefully at the same time wetting your appetite to decipher these magnificent symbols one day.

Complex functions could not have existed if it had not been for some high-level curiosity and experimentation. Ever since the time of Cardano and Tartaglia battling to solve higher degree polynomial equations (and reach their perceived

32 (Kayumov 2019).

finishing line), mathematicians flirted with the idea of *not* stopping when they were getting equations such as $x^2 = -1$. Instead, deciding to take a leap of faith, they enquired what would happen if they did square root negative numbers? School mathematics, especially up to the age of 16, tells us that when you multiply a negative by a negative you get a positive; this renders taking the square root of a negative number impossible! But who said mathematics does not need a bit a bravery from time to time? So, the solution to $x^2 = -1$ would be $\sqrt{-1}$. To square root any other negative number, it is sufficient to have the square root of -1 at hand, and simply multiply any of the "regular" results by that. For example, $\sqrt{-4} = 2i$.

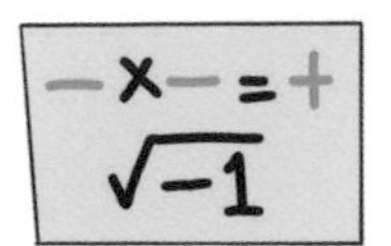

It would not be a spoiler to say that now, the imaginary unit $i = \sqrt{-1}$ is part of all sorts of calculations; you have already seen it in the chapter about De Moivre and his namesake formula! And remember, it was called "imaginary" in a derogatory manner, to say people using *i* had lost their minds and started *imagining* things; yet another struggle for new concepts to take on (but at least this time no one tried to throw anyone off a boat). Raphael Bombelli[33] used imaginary numbers in his book *Algebra* in the Sixteenth Century. A complex number is constructed

33 Raphael Bombelli (1526-1572) Italian mathematician.

using a Real number (any number that you have come across so far *knowingly* in your life is a Real number) as one part and an Imaginary number as the second part. For example, $2 + 3i$ or $5 - i$ are complex numbers. Even i on its own can be considered a complex number, with a real part equal to zero. So far so good. Or not quite? Having complex numbers unleashes all sorts of peculiarities. The first thing to spot is this: are complex numbers comparable in terms of size, like Real numbers?

Jean-Robert Argand[34] thought to use geometry to sort complex numbers somehow. Real numbers fit neatly on a number line. Ok, not quite so neatly as there are infinite fractions and *more infinite* decimals, but they fit. We know where to find them, and we know moving rightwards they *always* get bigger. But complex numbers will not abide to this rule. They have two parts, so should we compare part to part, and if yes which part would take precedence? There are no answers to these questions. Argand took the pair of axes and used the horizontal one to represent the real part and the vertical one to represent the imaginary part. So, we can find complex numbers on the 2D plane; we know where to find them but not how to order them. But that is ok, as they simply cannot be ordered.

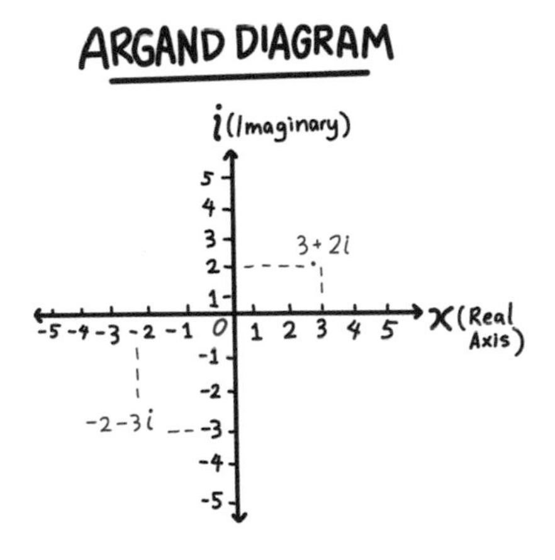

34 Jean-Robert Argand (1768-1822) Swiss mathematician

The position of each complex number on the Argand diagram can also be described by the distance of the point from the origin (called the modulus, symbolised by r) and the anticlockwise angle it forms with the positive real axis (called the argument or phase, symbolised with θ). The modulus-argument form can be written as $z = cos\theta\ isin\theta$ or more simply as $z = re^{i\theta}$.

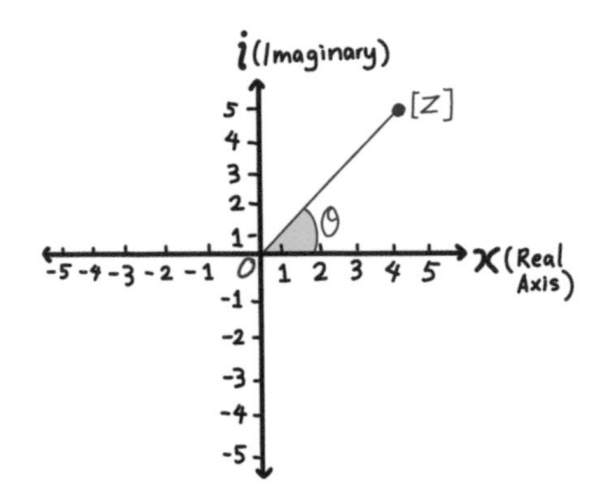

So, we now know what complex numbers are, and a couple of ways to depict them. Complex functions are mappings between two sets of complex numbers. But they are slightly trickier to depict than Real functions. In school, when you want to (ok, who am I kidding . . . when you are asked to) depict a function, you are merely asked to plot a graph. You take some values for the independent variable x, or the input, put it through the given function (for example $f(x) = x^3$, or $f(x) = \frac{1}{x}$). Then you obtain a set of outputs, or y values (dependent variable), you plot those points on the Cartesian set of axes and connect them. So, using *two* variables you draw a graph on a 2D plane.

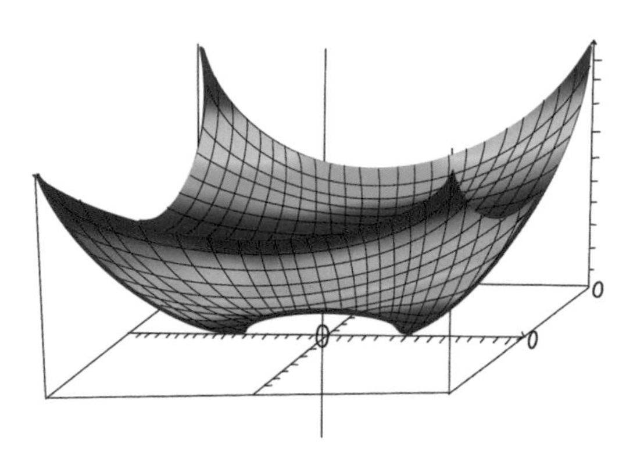

But complex functions map a combination of two values to a combination of two values. That gives essentially four variables, which poses, as you can *imagine*, particularly *complex* issues with depicting them in a 3D world. More recently, domain colouring, allows for the inputs to be plotted on a pair of axes. The output is depicted using a third axis to represent its modulus, and the argument is ingeniously represented by the hue. This representation of complex functions was not around when Bloch was working on them, but it is a great way to visualise complex functions.

THE DEATH OF BLOCH

Jacques Hadamard, one of his correspondent mathematicians, and editor of a journal, was overly impressed with the elegance of Bloch's writings and proofs. As he had not heard of Bloch before, he invited him to dinner to discuss a paper Bloch had submitted. Bloch could not possibly leave the asylum. So, he instead invited him over. Hadamard had great difficulties believing that Bloch was in a mental institution but decided to visit him regardless.

As Hadamard came to admire Bloch's work, he continued to visit repeatedly, and they both enjoyed engrossing mathematical exchanges. However, at some point, stumbling across an area they had both made significant strides in, Hadamard lost his temper. He accused Bloch of stealing his ideas and publishing some of his work under the pseudonym René Binaud. Hadamard strangled Bloch to death over the papers that he had been working on for the last week, before carers made it to Bloch's room.

AND NOW A BIT OF STATISTICS

Probability of being murdered because of jealousy: Most murders involve men (around 80%) and it is usually the domestic ones that have to do with jealousy, those victims usually being the female spouses.

Probability of death by strangulation: Homicide by asphyxiation is again more common amongst women but still a common cause of homicide death in general.

Probability of being killed while in a controlled/care environment: More likely to be killed by a so-called "angel of mercy", a carer/serial killer who finds you would be better off without the suffering, rather than by a visitor.

SELF-IMPOSED STARVATION AND OTHER DIFFICULTIES

KURT
GÖDEL
(1906-1978)

GÖDEL'S LIFE

Gödel was born in Brünn, in Austria-Hungary, an empire which was dissolved after defeat (with Germany) in the First World War. His hometown is now called Brno and is part of the Czech Republic. But during Gödel's childhood, this would have been what was then known as Czechoslovakia. In the duration of his life, he also became Austrian, German and American.

Gödel came from an affluent family. His father partly owned a textile company. His mother was well educated, and he kept close to her with frequent correspondence throughout his life. He seemed to have had a good childhood, aside from frequent childhood illnesses. This may have contributed in him developing a variety of health-related fears and general hypochondria for the rest of his life.

During his school years he demonstrated special interest in languages, religious studies, and mathematics. He attended university in Vienna, where he started studying physics, but quickly switched to mathematics and philosophy. Being taught by Hans Hahn[35] and Moritz Schlick[36] he became one of the youngest members of the Vienna circle.[37]

He obtained his doctorate at the same university in 1929 under Hans Hahn. In the decade that followed, Gödel produced some ground-breaking work, that will be discussed in the mathematics section to follow. In 1932 there was even a public lecture to present the new discoveries in mathematical logic, his own, and he not only bought a ticket to attend, but he also kept the ticket till the end of his life. He got invited to the Institute for Advanced Study at Princeton University for brief periods of time, but he would return to a devastated home country with the Nazi threat giving rise to fear and unrest. This, along with Schlick's murder by a former student cost him dearly emotionally and he entered sanatoriums intermittently during that decade.

Despite his family's objections, he got married to Adele Nimbursky Porkert in 1938. Adele was a former dancer and six years older than him. They were in a secret relationship for around ten years prior to the wedding, and Adele proved to be a supportive life partner to him.

35 Hans Hahn (1879 – 1934) Austrian Mathematician.

36 Moritz Schlick (1882-1936) German philosopher, Vienna Circle founder.

37 Between the years 1924 and 1936, social and natural scientists met regularly at the University of Vienna in what was called "The Vienna Circle of Logical Empiricism".

As the political scene in Europe continued shifting, and Gödel being (amongst other nationalities) German, he feared military conscription during the Second World War. Such an event would interrupt his already very high-level achievements. After having to overcome several bureaucratic obstacles, Gödel managed to arrange to flee for the US with his wife. He never returned to Vienna to see his mother, despite regular correspondence and promises he would do so; his fear of becoming trapped in Europe was insurmountable.

His wife and himself were both granted American citizenship less than a decade later. When studying for his citizenship exams he spotted a contradiction in the constitution of the United States, which he insisted on pointing out during his interview. His lawyer managed to stop him before bringing himself into a more difficult position and jeopardising his naturalisation.

In America, he worked at Princeton, initially as an ordinary member, progressing to a permanent member and eventually a professor. He became close friends with Albert Einstein,[38] often seen taking long walks together. Einstein expressed great respect for Gödel and a particular appreciation for their walks and talks.

38 Albert Einstein (1879-1955) German, Swiss *and* American theoretical physicist (maybe there was a citizenship Einstein held that you did not know about?).

GÖDEL'S MATHEMATICS

Gödel is most famous for proving two incompleteness theorems. In his doctoral thesis he proved the completeness theorem; all these lie within the realm of mathematical logic. An interpretation of the completeness theorem is that it offers connections between semantics and syntax and that if a formula is valid then it must be provable. So far so good. However, what was to come later though shook the mathematical community for ever. And a seemingly simple statement was used to make some interesting points:

"This statement cannot be proved."

In mathematical logic, statements can be either true or false. If this statement is false, it means that the statement can in fact be proven, which would make it true, and which obviously contradicts the assumption that it is false. If we instead assume that it is true, then great, but it cannot be proved so we would have to just accept this statement without a proof.

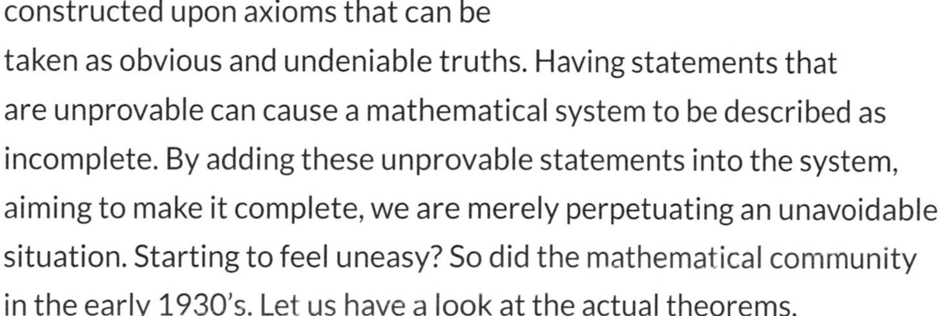

The entire mathematical edifice is constructed upon axioms that can be taken as obvious and undeniable truths. Having statements that are unprovable can cause a mathematical system to be described as incomplete. By adding these unprovable statements into the system, aiming to make it complete, we are merely perpetuating an unavoidable situation. Starting to feel uneasy? So did the mathematical community in the early 1930's. Let us have a look at the actual theorems.

Gödel's first incompleteness theorem: "any sufficiently expressive mathematical system must be either incomplete or inconsistent". Incompleteness means that there will be statements that could hold true, but they will never be proven. Inconsistency would cause much bigger problems; as the word suggests we would end up with mathematical statements opposing one another. Mathematics has traditionally been the rigid discipline people can trust and rely upon to always tell the truth. But with Gödel's result, we must accept that some things we will never know.

Gödel's second incompleteness theorem: "a consistent mathematical system cannot prove its own consistency". So even when we are lucky enough to work within a consistent system, we would not be able to know that. Not being able to know is an example of the incompleteness of that system.

These theorems were proven by Gödel just a few short decades after David Hilbert[39] produced a list with some famous mathematical problems awaiting solution. One of them was to prove that mathematics is consistent. The rest of those problems may or may not be provable, as we now know. And there are quite a few conjectures up for proof or refutation. And for some, a proof will never come.

So according to Gödel's theorems, some statements are true and not provable, and some are false and not refutable. But there is also a third category: the undecidable ones.

39 David Hilbert (1862-1943) German mathematician.

Georg Cantor[40] had shown, to the surprise of many of his contemporaries, that there are infinite different infinities. The smallest infinity is called the "countable infinity", being the one carried by the Natural Numbers {1,2,3, . . .}. Any collection of distinct objects (called a set as we saw in chapter 8) that is in one-to-one correspondence with the Natural Numbers also has the same infinity. We also know that the set of Real Numbers has an infinity greater than the countable one, called the continuum. What we do not know is whether there is an infinity between the countable and the continuum one. But does it matter?

Does it matter? Who would think we would be allowed to ask this question in mathematics, where it feels like everything is extremely strict *and* structured *and* important *and* does matter! Here is yet another surprise awaiting. Whether there is an infinity lying between countable and continuum does not matter; it is in fact undecidable.

What does undecidable mean? Gödel showed that no contradiction would arise if the continuum hypothesis were added to conventional Zermelo-Fraenkel set theory.[41] Paul Cohen[42] proved years later that no contradiction would arise if the *negation* of the continuum hypothesis were added to set theory. Hence it literally makes no difference whether the continuum hypothesis is true or false.

40 Georg Cantor (1845-1918) German mathematician.

41 Zermelo-Fraenkel set theory is a mathematical system set up in way to make sure that there will be no paradoxes arising within its rules. Whilst it sounds a bit too formal, within this theory you can perform all the arithmetic that you know without worrying that something weird will come up.

42 Paul Cohen (1934-2007) American mathematician.

Gödel's mathematics was ground-breaking, to the point that it changed the face of mathematics as it was known until then. There are still lots of old mathematics problems that have not been solved, and new work is produced every day. Incompleteness theorems make it more of a gamble to try and attack some of those problems than ever before. But maybe even more exciting.

THE DEATH OF GÖDEL

Despite his deep devotion to his work Gödel was becoming increasingly anxious and fearful, and the hypochondriac features of his character were manifesting frequently. He kept reducing his food intake in fear of someone wanting to poison him. His wife managed to persuade him to eat as she would taste his food before him or by being the only one allowed to prepare his food. When she had to be hospitalised for an extended period of time however, his condition continued to deteriorate further, to the point he starved himself to death. His weight eventually reached 29 kilograms. On his death certificate the cause of death recorded reads: "starvation and inanition, due to personality disorder." His wife outlived him by around three years.

AND NOW A BIT OF STATISTICS

Probability of dying of starvation: Several millions of deaths around the world per year can be so described. Severe acute malnutrition leads to the immune system being unable to fight infections. However this does not refer to intentional avoidance of food.

Probability of dying of an eating disorder: It is the highest amongst all mental illnesses, but women are more affected by it than men.

Probability of dying of fear: fears can often be irrational and cause anxiety rather have any more direct tangible consequence.

RESTORATION OF THE TRUTH

As mentioned in the introduction, one of these deaths is not true; but additional information has been provided in relation to all deaths for you to make an informed guess. Are you ready for the answer to be revealed? Please do not disclose this to people who have yet to finish reading this book – it will be a significant spoiler.

André Bloch was indeed conscripted along with his brother Georges to fight in World War I. Georges soon lost an eye in battle and was discharged. André, suffering less severe injuries (including falling from a watch tower) was kept on duty after a short recuperation period. Whilst recovering he was invited for dinner at his aunt's and uncle's house, with Georges attending that dinner too. André during the dinner stabbed all three family members to death. He initially run, but was arrested shortly after, without resistance. He was placed in the mental institution after this horrific event, where he stayed for over thirty years, producing mathematics throughout this period. He had used the pseudonym René Binaud, for fear of his Jewish heritage putting him in danger during World War II, and not to try and plagiarise (like suggested in Bloch's chapter). There was never such suspicion about his work.

Many years after the murders, during a chat with one of his doctors, Bloch revealed his reason for the murders. What he described was essentially eugenics – a belief system (or even worse a practice) that was fashionable at the time, but at the very least distasteful now. His delusion was that the mentally ill needed to be exterminated from society. He found that killing off a certain branch of his family was the logical thing to do. It was not, not least because one of the victims was not even a blood relative. And more generally, murder is hardly logical, or a solution to anything. Therefore, Bloch was a murderer himself, and he was not murdered. His recorded death was of leukaemia, at Saint Anne's Hospital, the only other place he had been to since entering the mental asylum.

Jacques Hadamard was simply a polite guest of Andre Bloch's, and his regular correspondent, who just got to know him quite well, not at all his murderer. In fact, Hadamard, wrote a book called *The Mathematician's Mind: The Psychology of Invention in the Mathematical Field*, describing how unconscious thoughts can emerge after lengthy conscious involvements with the problems at hand. There is no explicit mention of Bloch in his book, but Bloch's unusual story might as well have been part of what triggered Hadamard's curiosity to explore mathematical thinking at greater depth.

ANACHRONISMS AND SOURCES

Most of the illustrations in this book have one item or symbol from the wrong time period: these are "anachronisms". Here are those anachronisms – in italics – and brief notes about them.

How many did you get?

PYTHAGORAS 570-495 BC

- Stringed instruments have been around from millennia, starting with lyres possibly, but *guitars* only came about in 15th Century Spain.
- The first *black hole* was discovered in 1964, and it would probably be too far from anyone's imagination back in Pythagoras's time.
- *Nike of Samothrace* (Winged Victory of Samothrace) was created around 190 BC and could not have been sitting on Pythagoras's shelves (not least because it is 2.44m tall).
- *Hindu-Arabic numerals* were created sometime between the 6th and 7th century AC in the Hindu-Arabic world and soon evolved to the modern numerals that we use nowadays.

- Pilatre de Rozier trialled the first *hot air balloon* in 1783 with some animal passengers – not animal rights back then as it seems!

HIPPASUS 530-450 BC

- *Clefs* used as symbols representing the pitch of a musical piece were not invented until the 9th century.
- *Wristwatches* were not invented until 1868, and Achilles could not have had one.
- *Lightbulbs* were invented in 1879 by Thomas Edison.
- Although shovels made of animals' shoulder blades were used since the Neolithic age, *metal shovels* designed with specific tasks in mind were introduced much later.
- The first known *lighthouse* was Pharos of Alexandria built in 280BC.

ARCHIMEDES 287-212 BC

- This anachronism is not even our insertion. Archimedes using a *telescope* appeared in an advertisement for Yarwell's spectacles in 17th century London. Telescopes were invented in the 17th century, long after Archimedes' time.
- Our galaxy, the *Milky Way* was of course there in Archimedes's time too! But he did not know that. The Milky Way was discovered by Galileo Galilei in 1610.
- *Rubber ducks* were created in the 1800s when rubber became soft and malleable enough. Originally, they were solid and could not float though!
- A cutting apparatus resembling shears appeared in Italy in 400BC, but it took another two to three hundred years before there was fitting around the fingers resulting in what we now recognise as *scissors.*

- Hot baths have been enjoyed for millennia, *bubble baths* however are more of a 20th century invention.
- The *pirate flag*, white skull and crossbones on a black background, was first seen in the 1700s, so Archimedes was not defending against pirate ships.
- Pies and jams were around for quite a while, but *recipes* appeared in the first ever cookbook, *De Re Coquinaria*, in the 4th century AD.
- *Tablecloths* started being used in Europe in the 1st century AD, so Archimedes would not have had his table covered in one.
- In a world littered with plastic, it is difficult to think that it has only been around for a century and a bit; *plastic* was invented in 1907 and it contains no molecules found in nature. So, there would not be any beach toy buckets near where Archimedes was working.

HYPATIA 360-415 AD

- Codex manuscripts had just started appearing in the 4th century, and Hypatia might as well have had access to them. However, she could not have had a hardbound book with a *colourful cover. ISBN* numbers (International Standard Book Number) were introduced in 1967.
- Whilst compass and straightedge were essential in geometric construction since Ancient Greece, the *protractor* is first mentioned in writing in 1589 by Thomas Blundeville – it was used for map reading.
- Sanitary pads only acquired *wings* in the last 2-3 decades to stop them from moving; in Hypatia's time women would use thick rags.
- *High heels* were originally created to make it easier for men to ride on horses and the first record of a woman wearing heels was in the 16th century.
- *Graduation hats* (mortarboards) are first seen in photos of graduates in the 1800s.

- Witchcraft was not originally depicted using *black pointy hats*; they became a staple of witches in the early 18th century through illustrations in children's books. *Brooms* came a bit earlier in the 15th century.

GEROLAMO GARDANO 1501-1576

- A version of the green-white-red flag has been in use since 1797 and has been the official *flag of Italy* since 1948.
- Bartolomeo Cristofori invented *the piano* around 1700 as an improvement to the harpsichord.
- Thermoscopes, the predecessor of a *thermometer* was invented in 1593 by Galileo Galilei. It has been improved since several times, reaching the digital, contact-free ones today.
- The first *stethoscope* was invented in 1816 by Rene Theophile Hyacinthe Laennec, and (as pictured) it could only be used with one ear (to hear the heart and lungs), unlike more modern versions.
- Tartaglia was also a topographer; the map here however is rather modern and depicts *Australia*, which was not known to the Europeans until 1606.
- *Raising one's right hand* when taking an oath was established in the 17th century London.
- The idea of projecting images significantly predates photography; the first *camera* that could be reproduced to be sold was created in 1839 by Alphonse Giroux.
- *UNO playing cards* were invented in 1971 and they are not particularly associated with gambling. Playing cards were invented in the 9th century China and reached Europe by the 14th century.
- The *dollar sign* was created in the 1790s in Philadelphia.
- Religion has had a prime role in Cardano's life, however Quakers

were not formed until the middle of the 17th century, and Cardano could not have known their *black/red star symbol.*

TYCHO BRAHE 1546-1601

- *Racing bicycles* are a Twentieth Century invention.
- *Champagne* was invented in 1697 by monk Dom Perignon. It now enjoys Protected Designation of Origin (PDO).
- *Toy balloons* became popular in the late 1800s; the idea was originally conceived by Michael Faraday to help with laboratory experiments.
- Toilet roll was not commercially available until 1857.

ABRAHAM DE MOIVRE 1667-1754

- Alexander Graham Bell invented and patented the *telephone* in 1876-77.
- The *three-point seatbelt* was introduced in 1959 and replaced the less safe two-point seatbelt that was just a few years older. Nils Bohlin was the engineer behind this invention.
- *Paper clips* were invented in 1899 and have stayed the same through the years.
- *Safety pins* used as clothes fastenings or other purposes where invented in 1849 by Walter Hunt.
- There were no postal boxes on the streets in the UK until 1840. Nowadays practically everyone in the UK has a *postal box* within walking distance. They were all red by 1884.
- Whilst it is very likely people were chewing resin since pre-historic times, chicle gum was patented 1869 in the US who is currently the largest producer and exporter of *chewing gum.*

- Whilst gambling has been around for an exceptionally long time, the bets would be placed mostly using anything made of gold; round *poker chips*, made from a variety of materials, clay, wood, ivory etc appeared in the mid to late 1800s. Some materials allow for many variations in shape and colour.
- Some writing apparatus started emerging as early as the 14th century, with many tweaks and improvements made over the following centuries. Although usable, the *typewriter* did not reach its well-known functionality and appearance until 1868.
- *Matches* were discovered accidentally in 1826 by John Walker who was experimenting with a paste that caused the stick it was stuck to, to catch fire after friction.

EVARISTE GALOIS 1811-1832

- *Sellotape* was invented in 1937, and the "c" from cellophane was changed into an "s" so that it could be trademarked.
- *Glue sticks* were invented in 1969 and they made use of a mechanism used for lipsticks by Guerlain almost a century earlier.
- *Cars* with internal combustion engines were invented in 1885 by Karl Benz.
- *Mirrors* were invented in 1835 by Justus von Liebig, using a layer of metallic silver on glass. Other types of reflecting surfaces have been around for much longer.
- There would be no *ball point pens* available to Galois while writing up his findings; fountain pens made an appearance in 1827, relieving some of the frustration having to keep dipping the quill or metal tip pen in ink. However, ball point pens emerged at around 1888 and offered significant advantages.

- The *French flag* has a bit of a turbulent story from the late 18th to the mid-19th centuries, with a Royal Navy white flag sometimes being used and symbolisms of the colours coming and going. The blue-white-red became the official flag of France in 1848.
- *Electric doorbells* were invented in 1831 but relied on expensive batteries. This stopped them becoming popular until 1913 when they could be energised through the electric current of the household.
- *Martini* was probably created in the late 1800s; more than one stories are conflicting with regards its exact first creation. But most certainly, not earlier than 1880.
- Galois opponent no matter how wealthy and in vogue would not have access to *denim*, patented by Levi Strauss and Jacob Davis in 1873. *Skinny jeans* are a 2005 invention and as the years go by, they become ever more skin-tight.

ANDRE BLOCH 1893-1948

- *Zip fasteners* were invented in 1913 and had they taken less time to be perfected and mass produced, they could have been found on military uniforms in WW1 too.
- *Intravenous dripping devices* were first used in medicine in 1902 during a cholera epidemic, but actually were not widely used before 1950.
- Whilst there are several early versions of *wheelchairs*, a *foldable* one with metal rims to manoeuvre were not available before the 1950s.
- *Floppy discs* were discovered in 1971, so no storage memory for Bloch other than paper.
- *Radio broadcasts* were started in France in 1923, not that if they were already around Bloch would be at all interested!

KURT GÖDEL 1906-1978

- *CMI (Clay Mathematics Institute)* was founded in 1998 to acknowledge and promote the value of mathematics, development and dissemination of mathematical knowledge.
- *Post-it notes* took a while to be fully developed by 3M and hit the market in 1980 (or late 1979) and have been a great success since.
- Gödel feared many microorganisms causing all sorts of diseases, but *SARS-CoV-2*, the virus causing COVID-19 emerged, well, in 2019.
- *Maryam Mirzakhani* was the first woman to even win the Fields Medal in 2014. Her work was on Riemann surfaces and their moduli spaces. One of the thinking clouds over Gödel's head contains her way of visualising *her area of research*.
- Some photographic evidence suggests that on occasions women would use luggage on wheels, but the actual *rolling luggage* was not invented until 1972. Plus, women would have been unlikely to carry their own luggage in the first half of the 20th century.
- Although the experimentation with *contact lenses* started in the 19th century, they were not perfected, licensed and available to purchase until 1971; hence Gödel should have been wearing his glasses during his long walks with Einstein.
- *Cronuts* were invented in 2013 by a French baker working in New York; they are a combination of croissants and doughnuts.
- Headphones were invented in 1910 by Nathaniel Baldwin so the anachronism here is the *smartphone*, invented in 1992.

SOURCES

Online Sources for these can be found on the website at

https://www.tarquingroup.com/anachronisms